T0199256

Last *of the* Curlews

Last *of the*

Curlews

Fred Bodsworth

Foreword by W. S. Merwin
Afterword by Murray Gell-Mann

*Illustrated by Abigail Rorer
based on original drawings by T. M. Shortt*

COUNTERPOINT
WASHINGTON, D.C.

LAST OF THE CURLEWS was originally published in 1955 by Dodd, Mead & Company.

FIRST COUNTERPOINT EDITION, designed by David Bullen.

This edition is printed in the United States of America on acid-free paper that meets the American National Standards Institute Z39–48 Standard.

Library of Congress Cataloging-in-Publication Data
Bodsworth, Fred, 1918–
    Last of the curlews / Fred Bodsworth ; foreword by W. S. Merwin ; afterword by Murray Gell-Mann ; illustrated by Abigail Rorer.
        p.   cm.
    "Based on original drawings by T. M. Shortt."
    Includes bibliographical references.
    1. Eskimo curlew—Fiction.   2. Birds—Migration—Fiction.   3. Extinct birds—Fiction.   I. Title.
PR9199.3.B56L37   1995
813'.54—dc20   95-33194

ISBN 978-1-5824-3735-4

C O U N T E R P O I N T
P.O. Box 65793
Washington, D.C. 20035-5793

Typeset by Wilsted & Taylor
Distributed by Publishers Group West

10 9 8 7 6 5 4 3 2

PERSEUS
POD
ON DEMAND

Contents

Foreword

W. S. Merwin

Anyone to whom books are of interest is familiar with the curiosity they awaken when one sees them in other people's houses, lining the walls or stacked precariously so that the top ones have to be moved in order to inspect what is underneath. A hint of sinful indulgence, akin to memories of skipping school and the imaginary taste of forbidden fruit, haunts such inquisitiveness, even under roofs where the visitor has been warmly welcomed and made to feel at home. The less accustomed to their place the books seem to be, the more intriguing they appear, as though they might be about to leave again before they had been noticed. They are survivors when others of their race have vanished all around them.

In the late summer of 1994 I was visiting friends who had recently moved to the island of Kauai. There were still stacks of books in the entrance room off the garage, and in the rooms where I was staying the shelves were only partly filled. The books there, I realized, had been sifted again and again. I came to a small paperback, cheaply produced and of undistinguished appearance: *Last of the Curlews,* by Fred Bodsworth, with drawings (by T. M. Shortt) that reminded me of an edition of Aldo Leopold's *Sand County Almanac.*

I could not tell at a glance how serious a book it might be, what kind of reader it had been intended for. I wondered whether the title referred to the Eskimo curlew, one of the creatures that I had learned I was too late to see, for it had become extinct in my own lifetime while I was a child in a city turning the pages of books with pictures of birds. All of the migrants, and the phenomenon of migration itself, from the time I first knew of them, seemed to possess a dimension and radiance of their own. One of the things that we late arrivals on the earth sense, and in turn portray as legendary, is an awareness much older than our own. In each of the migrants we are afforded glimpses of beings in whom the turning of the earth, its axis and magnetic fields, day and night and the seasons, the winds and all that we call the elements, and the motions of the celestial bodies themselves are the pulse of a heart and the illumination of a consciousness.

Where I live, in Hawaii, in the latter part of the century, the Pacific golden plover is such an emanation. When the plovers return from Alaska in August, the rising call notes as the birds flash overhead are like the touch of a wand. There is no land between the Alaskan coast and here, and those wings have covered the distance in a single flight. All winter the plovers streak above the slopes here in the late afternoon or at night in the moonlight, calling the same call they sent into the sky over the huge empty sea. In the morning I see a single plover on the open grassland or along the same section of dirt road, standing still or walking, like a shadow. When I realize in the spring that the plovers have gone, there is always a breath of fear in their place—from a ghost not theirs, of course, but my own. Among the migrants, those that will not return inhabit a farther eminence, and the Eskimo curlew is high among those. Its route, I had learned, had been the most enormous trajectory of all: from the arctic to the antarctic every autumn, and in the spring the entire transit north again. The curlews' shape, their wings and feathers, had been incomparably elegant, their flight swift and tireless, and their vast flocks had been obliterated without hesitation in the course of little more than a century by the greed and indifference of a species that considers its own withering arrival an improvement and that managed to believe it was winning the West.

Their destruction was part of a process that has been accelerating in a sort of geometrical progression all through our age until the rate of species extinction has become a regular topic in the news. From time to time, with a frequency that is hard to assess, the figure of the last representative of some kind of life, some way of being—a palm, a rhinoceros, a sparrow, a fish, a word from a language, a speaker of a language—presents itself to us for a moment like a recurring dream in another form, and then goes out, leaving us to make of it what we can. Whether we acknowledge it or not, this last figure, in whatever guise it returns, is a single irreducible warning, and we know it even as we continue to organize our restlessness and our demands around the assumption of our own unending growth.

I opened the book.

I was standing in the living room, which was filled with afternoon light. Mr. Bodsworth wrote with plain, succinct evocation and beauty of the arctic autumn and the dwindling of the once enormous assemblies of curlews that for so long had gathered at that season to begin the journey south. During the past century, while the extinctions of species have continued to pick up speed, we have been taught to eschew the delusions of anthropomorphism, the temptation to project our own evaluations, responses, and personalities onto the actions and gestures of other forms of life. It has been an essential and complex cleansing and is never likely to be complete. But in the usage of some it has been turned into a new superstition and obscurantism, a tempered reinforcement of human arrogations, and a justification for denying any continuity between the consciousness and sensations of our kind and all the rest of life. Some part of the suffering of each creature is incommunicable and solitary. I cannot believe that humans are unique in this, and I am convinced that when we forget it we mistake ourselves. Mr. Bodsworth is a scientist, a trained observer of natural phenomena from the viewpoint of post-eighteenth-century biology with its creed of objectivity, and he was obviously determined to avoid, in his writing, the specter of anthropomorphism. But he was describing behavior and the expression of overpowering urges, drives, passions, and he had only our own language with which to do it. Besides, it is obvious that he cared about the fate of the protagonist of his narrative. Making use of all the knowledge and reasonable probability available to him, his book is an articulate ornithologist's account of how the migratory flight of the last, or one of the last, of the Eskimo curlews might have taken place.

I was caught from the beginning, and for the rest of the visit I returned to the story whenever I could. I had not managed to finish it by the time I was to leave, and I was not sure of being able to find another copy. My hosts lent me theirs, and I came to the last page on the way home, in the air.

The account seemed to me a classic and I wondered how I had managed to miss it. I learned almost at once that Jack Shoemaker, after running North Point Press for a decade, was starting a new publishing house, and I told him about it. He and his assistant managed, with some difficulty, to get a copy from the Library of Congress, and their response to the narrative paralleled mine. They sought out the author and began plans for the present edition.

None of us, at that point, knew of Fred Bodsworth's continuing attention to the curlew since he finished the book, and the cautious, fragile hope that he had to offer in his epilogue. And of how we would be left, for the time being, clinging.

March 1995

The beauty and genius of a work of art may be
reconceived, though its first material expression be
destroyed; a vanished harmony may yet again
inspire the composer; but when the last individual of
a race of living beings breathes no more, another
heaven and another earth must pass before such a
one can be again.

From THE BIRD: ITS FORM AND FUNCTION,
by C. William Beebe

Last *of the* Curlews

Chapter One

By June the arctic night has dwindled to a brief interval of grey dusk and throughout the long days mosquitoes swarm up like clouds of smoke from the potholes of the thawing tundra. It was then that the Eskimos once waited for the soft, tremulous, far-carrying chatter of the Eskimo curlew flocks and the promise of tender flesh this chatter brought to the arctic land. But the great flocks no longer come. Even the memory of them is gone and only the legends remain. For the Eskimo curlew,

originally one of the continent's most abundant game
birds, flew a gantlet of shot each spring and fall, and,
flying it, learned too slowly the fear of the hunter's gun
that was the essential of survival. Now the species lin-
gers on precariously at extinction's lip.

The odd survivor still flies the long and perilous
migration from the wintering grounds of Argentine's
Patagonia, to seek a mate of its kind on the sodden tun-
dra plains that slope to the arctic sea. But the arctic is
vast. Usually they seek in vain. The last of a dying
race, they now fly alone.

As the arctic half-night dissolved suddenly in the pink and then the glaring yellow of the onrushing June day, the Eskimo curlew recognized at last the familiar S twist of the ice-hemmed river half a mile below. In the five hundred miles of flat and featureless tundra he had flown over that night, there had been many rivers with many twists identical to this one, yet the curlew knew that now he was home. He was tired. The brown barbs of his wing feathers were frayed and ragged from the migration flight that had started in easy stages below the tropics and had ended now in a frantic, nonstop dash across the treeless barren grounds as the full frenzy of the mating madness gripped him.

The curlew set his wings and dropped stonelike in a series of zigzagging sideslips. The rosy pink reflections of ice pans on the brown river rushed up toward him. Then he leveled off into a long glide that brought him to earth on the oozy shore of a snow-water puddle well back from the riverbank.

Here for millenniums the Eskimo curlew males had come with the Junetime spring to claim their individual nesting plots. Here on the stark arctic tundra they waited feverishly for the females to come seeking their mates of the year. As they waited, each male vented the febrile passion of the breeding time by fighting savagely with neighboring males in defense of the territory he had chosen. In the ecstasy of homecoming, the curlew now hardly remembered that for three summers past he had been mysteriously alone and the mating fire within him had burned itself out unquenched each time as the lonely weeks passed and, inexplicably, no female had come.

The curlew's instinct-dominated brain didn't know or didn't ask why.

He had been flying ten hours without stop but now his body craved food more than rest, for the rapid heartbeat and metabolism that had kept his powerful wing muscles flexing rhythmically hour after hour had taken a heavy toll of body fuel. He began probing into the mud with his long bill. It was a strange bill, curiously adapted for this manner of feeding, two and a half inches long, strikingly downcurved, almost sicklelike. At each probe the curlew opened his bill slightly and moved the sensitive tip in tiny circular motions through the mud as he felt for the soft-bodied larvae of water insects and crustaceans. The bill jabbed in and out of the ooze with a rapid sewing-machine action.

There were still dirty grey snowdrifts in the tundra hollows but the sun was hot and the flat arctic world already teemed with life. Feeding was good, and the curlew fed without stopping for over an hour until his distended crop at the base of his throat bulged grotesquely. Then he dozed fitfully in a half-sleep, standing on one leg, the other leg folded up close to his belly, his neck twisted so that the bill was tucked deeply into the feathers of his back. It was rest, but it wasn't sleep, for the curlew's ears and his one outside eye maintained an unrelaxing vigil for arctic foxes or the phantomlike approach of a snowy owl. His body processes were rapid and in half an hour the energy loss of his ten-hour flight was replenished. He was fully rested.

The arctic summer would be short and there would be much to do when the female came. The curlew flew to a rocky ridge that rose about three feet above the surrounding tundra, alighted and looked about him. It was a harsh, bare land to have flown nine thousand miles to reach. Its harshness lay in its emptiness, for above all else it was an empty land. The trees that survived the gales and cold of the long winters were creeping deformities of birch and willow, which after decades of snail-paced growth had struggled no more than a foot or two high. The timberline where the trees of the subarctic spruce forests petered out and the tundra Barren Grounds began was five hundred miles south. It was mostly a flat and undrained land laced with muskeg ponds so closely packed that now, with the spring, it was half hidden by water. The low gravel humps and rock ridges that kept the potholes of water from merging into a vast, shallow sea were covered with dense mats of grey reindeer moss and lichen, now rapidly turning to green. A few inches below lay frost as rigid as battleship steel, the land's foundation that never melted.

The curlew took off, climbed slowly, and methodically circled and recircled the two-acre patchwork of water and moss that he intended to claim as his exclusive territory. Occasionally, sailing slowly on set, motionless wings, he would utter the soft, rolling whistle of his mating song. There was nothing of joy in the song. It was a warcry, a warning to all who could hear that the territory had an owner now, an owner flushed with the heat of the mating time who would defend it unflinchingly for the female that would come.

The curlew knew every rock, gravel bar, puddle and bush of his territory, despite the fact that in its harsh emptiness there wasn't a thing that stood out sufficiently to be called a landmark. The territory's western and northern boundaries were the top of the river's S twist that the curlew had spotted from the air. There was nothing of prominence to mark the other boundaries, only a few scattered granite boulders that sparkled with specks of pyrite and mica, a half dozen birch and willow shrubs and a few twisting necks of brown water. But the curlew knew within a few feet where his territory ended. Well in toward the centre was a low ridge of cobblestone so well drained and dry that, in the ten thousand years since the ice age glaciers had passed, the mosses and lichens had never been able to establish themselves. At the foot of this parched, stony bar where drainage water from above collected, the moss and lichen mat was thick and luxuriant. Here the female would select her nesting site. In the top of a moss hummock she would fashion out a shallow, saucerlike depression, line it haphazardly with a few crisp leaves and grasses, and lay her four olive-brown eggs.

The curlew circled higher and higher, his mating song becoming sharper and more frequent. Suddenly the phrases of the song were tumbled together into a loud, excited, whistling rattle. Far upriver, a brown speck against the mottled grey and blue sky, another bird was winging northward, and the curlew had recognized it already as another curlew.

He waited within the borders of his territory, flying in tightening circles and calling excitedly as the other bird came nearer. The female was coming. The three empty summers that the male had waited vainly and alone on his breeding territory were a vague, tormenting memory, now almost lost in a brain so keenly keyed to instinctive responses that there was little capacity for conscious thought or memory. Instinct took full control now as the curlew spiraled high into the air in his courtship flight, his wings fluttering mothlike instead of sweeping the air with the deep strokes of normal flight. At the zenith of the spiral his wings closed and the bird plunged earthward in a whistling dive, leveled off a few feet above the tundra and spiraled upward again.

The other bird heard the male's frenzied calling, changed

flight direction and came swiftly toward him. But instinctively obeying the territorial law that all birds recognize, she came to earth and perched on a moss-crowned boulder well outside the male's territory.

The male was seething now with passion and excitement. He performed several more courtship flights in rapid succession, spiraling noisily upward each time until almost out of sight, then plunging earthward in a dive that barely missed the ground. For several minutes the female nonchalantly preened her wing feathers, oblivious to the love display. Then, alternately flying and running across the tundra a few quick wingbeats or steps at a time, she moved into the mating territory and crouched submissively, close to where the male was performing.

The male whistled shrilly and zoomed up in a final nuptial flight, hovered in midair high above the crouching female, then dropped like a falling meteorite to a spot about six feet from where she waited. He stood for a moment, feathers fluffed out and neck outstretched, then walked stiff-legged toward her.

When still a yard away, the male abruptly stopped. The whispering courtship twitter that had been coming from deep in his throat suddenly silenced, and a quick series of alarm notes came instead. The female's behavior also suddenly changed. No longer meekly submissive, she was on her feet and stepping quickly away.

The male abandoned his courtship stance, lowered his head like a fighting cock and dashed at the female. She dodged sideways, and took wing. The male flew in pursuit, calling noisily and striking repeatedly at her retreating back.

The curlew's mating passion had suddenly turned into an aggressive call to battle. The female was a trespasser on his territory, not a prospective mate, for at close range he had recognized the darker plumage and eccentric posture of a species other than his own. The other bird was a female of the closely related Hudsonian* species, but the Eskimo curlew knew only, through the instinctive intuition set up by nature to prevent infertile matings between different species, that this bird was not the mate he awaited.

He chased her a quarter of a mile with a fury as passionate as his love had been a few seconds before. Then he returned to the territory and resumed the wait for the female of his own kind that must soon come.

Two curlew species, among the longest-legged and

*The Hudsonian curlew is now commonly known as the whimbrel.
—The Author

longest-billed in the big shorebird family of snipes, sandpipers, and plovers to which they belong, nest on the arctic tundra—the Eskimo curlew and the commoner and slightly larger Hudsonian. Though distinct species, they are almost indistinguishable in appearance.

The arctic day was long, and despite the tundra gales that whistled endlessly across the unobstructed land the day was hot and humid. The curlew alternately probed the mudflats for food and patrolled his territory, and all the time he watched the land's flat horizons with eyes that never relaxed. Near midday a rough-legged hawk appeared far to the north, methodically circling back and forth across the river and diving earthward now and then on a lemming that incautiously showed itself among the reindeer moss. The curlew eyed the hawk apprehensively as the big hunter's circling brought it slowly upriver toward the curlew's territory. Finally the rough-leg crossed the territory boundary unmarked on the ground but sharply defined in the curlew's brain. The curlew took off in rapid pursuit, his long wings stroking the air deeply and his larynx shrieking a sharp piping alarm as he closed in on the intruder with a body weight ten times his own. For a few seconds the hawk ignored the threatened attack, then turned back northward without an attempt at battle. It could have killed the curlew with one grasp of its talons, but it was a killer only when it needed food, and it gave ground willingly before a bird so maddened with the fire of the mating time.

The sun dipped low, barely passing from view, and the curlew's first arctic night dropped like a grey mist around him. The tundra cooled quickly, and as it cooled the gale that had howled all day suddenly died. Dusk, but not darkness, followed.

The curlew was drawn by an instinctive urge he felt but didn't understand to the dry ridge of cobblestone with the thick mat of reindeer moss at its base where the nest would be. In his fifth summer now, he had never seen a nest or even a female of his kind except the nest and mother he had briefly known in his own nestling stage, yet the know-how of courtship and nesting was there, unlearned, like a carry-over from another life he had lived. And he dozed now on one leg, bill tucked under the feathers of his back, beside the gravel bar which awaited the nest that the bird's instinct said there had to be.

Tomorrow or the next day the female would come, for the brief annual cycle of life in the arctic left time for no delays.

The Gantlet

PHILOSOPHICAL TRANSACTIONS *of*
THE ROYAL SOCIETY *of* LONDON
giving some ACCOUNT *of the present
undertakings, studies and labours
of the* INGENIOUS *in many
considerable parts of the world.*
VOL. LXII *for the year 1772.*

Article xxix.

*An account of the birds sent from Hudson's Bay;
with observations relative to their natural history; and
Latin descriptions of some of the most uncommon. By
Mr. J. Reinhold Forster, F.R.S.*

*From the factory at Hudson's Bay, the Royal Society
were favoured with a large collection of uncommon
quadrupeds, birds, fishes, &c., together with some
account of their names, place of abode, manner of life,
uses, by Mr. Graham, a gentleman belonging to the set-*

tlement on *Severn River; and the governors of the Hudson's Bay Company have most obligingly sent orders, that these communications should be from time to time continued.*

(The birds described by Mr. Forster being all introduced into Mr. Latham's ornithological volumes, under the same titles, it becomes unnecessary to give the Latin descriptions, which are therefore omitted from these transactions.)

1. Falco Columbarius. *Pigeon hawk. It is migratory. 2 . . . . 3 . . . . 4. . . .*

18. New species. Scolopax Borealis. *Eskimaux curlew. This species of curlew, is not yet known to the ornithologists; the first mention made of it is in the* Faunula Americae Septentrionalis, *or* Catalogue of North American Animals. *It is called Wee-kee-me-nase-su, by the natives; feeds on swamps, worms, grubs, &c., visits Albany Fort in April or beginning of May; breeds to the northward, returns in August, and goes away southward again the latter end of September in enormous flocks.*

Chapter Two

The hot days and chilling nights raced by, the snowdrifts disappeared even from the shaded hollows, the austere browns and greys of the tundra became a flaming carpet of pink and yellow blooms, and the female curlew never came. Other shorebirds came in their hundreds, fought for their territories, mated, nested and prepared to bring forth the new cycle of life they had flown six or eight thousand miles to create. The male curlew fought insanely with every plover and sandpiper that crossed his territory boundary until the outer perimeters were flecked with the brown feathers of trespassers that had retreated too slowly before the curlew's onslaughts. The mating hormones poured out by his glands could only dam up within him like an explosive charge.

Instinctively the curlew fought every other shorebird that ventured near, yet in his instinctive behavior pattern there was no enmity for the buntings, longspurs and ptarmigans that also occupied the tundra—birds not biologically related and not competitors for the same insect food he would need for his own nestlings when the female came. When a female willow ptarmigan built her nest and laid her twelve buff eggs less than fifteen feet from the moss hummock where the curlew's nest would be, the curlew ignored her and in a few days forgot she was there.

The nights grew darker and longer. The tiny, brilliant flowers of the tundra dried into wisps of silk-plumed seed. Close by, a pair of golden plover, their black bellies and breasts glistening in the low rays of a morning sun, began calling excitedly and flying in rapid circles. The curlew knew their young had come, and like the young of all shorebirds, already well developed at birth, they had left the nest and were running about before the shells of the eggs that had held them were dry. The arctic summer was waning.

Several of the plovers' down-covered young scampered into the curlew's territory and the mother followed them with food. The curlew whistled a warning and flew toward her. But the call of her young was stronger than the fear of another much larger bird, and the plover stood her ground, her wings spread protectively before the tiny, peeping balls of yellow fluff that squirmed downward into the mat of reindeer moss. The curlew swerved upward without striking her. And he didn't attack again. Instead he circled to a rocky hummock a hundred feet away, alighted, watched the plover feeding her young for a moment or two, and then forgot her.

Within the curlew the annual rhythm of glandular activity had passed its peak and begun to ebb, and its product, the belligerent drive of the mating time, was dying. A new urge was replacing it. Where before, defense of the territory was an overriding demand that took priority over even the search for food, the curlew was now feeling the first stirrings of a restless call to move. No female had come. The territory was losing its meaning.

Periodically the curlew flew back at the golden plover, but

when the plover refused to fly the curlew would lose interest and forget her again. This went on for most of a day, the curlew suddenly remembering that there were intruders on his territory, then just as suddenly forgetting them. The next day other shorebirds moved in and out of the territory. Now the curlew ignored them. Once he flew far down river and was gone a couple of hours, the first time he had left the territory since arriving almost two months before.

Around him the young shorebirds of the year were maturing rapidly and their parents were abandoning them to fend for themselves. The disassociation between parents and young was abrupt and complete, the parents forming their own flocks, the young birds theirs.

It was late July. The tundra potholes and their muddy edges were teeming with the water insects and crustaceans on which the shorebirds fed. Food was at its peak of abundance and winter was still a couple of months away, but the arctic had served its purpose and now the distant southland was calling the shorebird flocks, many weeks before there was any real need for them to leave. The curlew who had fought savagely all summer to be alone, now felt a pressing desire for companionship.

There was no reasoning or intelligence involved. The curlew was merely responding in the ages-old pattern of his race to the changing cycle of physiological controls within him. As days shortened the decreasing sunlight reduced the activity of the bird's pituitary gland. The pituitary secretion was the trigger that kept the reproductive glands pouring sex hormones into the bloodstream, and as the production of sex hormones decreased, the bird's aggressive mating urge disappeared and the migratory urge replaced it. It was entirely a physiological process. The curlew didn't know that winter was coming again to the arctic and that insect eaters must starve if they remained. He knew only that once again an irresistible inner force was pressing him to move.

But somewhere in his tiny, rudimentary brain the simple beginnings of a reasoning process were starting. Why was he always alone? When the rabid fire of the mating time burned fiercely in every cell, where were the females of his species that the curlew's instinct promised springtime after springtime? And now with the time for the flocking come, why in the myriads of shorebirds and other curlews were there none of the smaller and lighter-brown curlews he could recognize as his own kind?

A few days later the lure of the territory disappeared entirely and the curlew rose high and flew southward for a couple of hours without alighting. He came down finally to feed on a small mudflat where a river emptied into a large lake. The tundra was now disgorging its summer population of shore-birds and flock after flock of southward moving sandpipers passed by. One flock of long-legged shorebirds, flying in a wavering V, swept low along the lakeshore. The curlew stopped his feeding and called excitedly, for the flight pattern and flock formation could only be curlew. The flock wheeled without breaking formation, moving with the precision and instantaneous timing of a single organism, as though one nerve centre controlled the movements of every bird. On stiffened, downcurved wings they glided in to the mudflat. The Eskimo curlew ran toward them, then stopped abruptly after a few strides and nonchalantly resumed his feeding. They were Hudsonian curlews with the shorter bills and buffy underparts that marked them as birds of this nesting.

The curlew didn't know that this other species, almost identical outwardly, was a slower flying bird unsuited as a migration companion. He didn't know that young shorebirds of the year develop their full wing strength late and are left behind by the adults to follow by instinct the perilous eight-thousand-mile southward route they have never seen before. His instinctive behavior code, planted deep in his brain by the genes of countless generations, told him only what do, with-

out telling him why. His behavior was controlled not by mental decisions but by instinctive responses to the stimuli around him. He desired the association of a flock, but the Hudsonians had failed to release the flocking response in his inner brain and now he ignored them in his feeding. When they flew again a short time later, the Eskimo curlew hardly noted their departure. In a land pulsing with the wingbeats of migrating shorebirds, the curlew was alone again.

By afternoon the mudflat was dotted with the darting forms of shorebirds that had stopped to feed. Most of them kept together in flocks of their own species. At dusk the flocks ceased feeding and took off, one by one, until only the curlew remained, the birds of each flock whistling sibilantly to each other to retain formation in the falling darkness.

They circled high until a half mile or so above the tundra, then leveled off and headed southward. It was usual for the shorebirds to migrate principally at night, for their digestion and energy consumption were rapid and the daylight was required for feeding. The high level of energy that migration demanded could be maintained only by timing the flights so that they ended with the dawn when feeding could be at once resumed.

Far above him, the curlew could hear the faint, lisping notes of the arctic migrants pouring south to a warmer land. Needles of ice began forming at the shallow edges of the mudflat puddles. The bird's instinct rebelled at the idea of flying alone, yet when he called shrilly into the cold night there was no answer, and the time had come when he had to move.

He turned into the breeze, held his wings extended outward and adjusted the angle—leading edge up and trailing edge down—until he could feel the lifting pressure of the wind beneath them. Of all the shorebirds' wings, the Eskimo curlew's—long, narrow and gracefully pointed—were best adapted for easy, high-speed flight. Even standing motionless with wings extended in the faint, night breeze, the bird was weightless and almost airborne. He pushed off gently with his legs, took a few rapid wingbeats with the flight feathers twisted so that they bit solidly into the air, and rose effortlessly. He climbed sharply for more than a minute until the tundra almost vanished in the grey dark below, then he leveled off and picked up speed with a slower, easier wingbeat. The air rushed past him, pressing his body feathers tightly against the skin. The migration had begun. Even the curlew's simple brain sensed vaguely that the unmarked flyway ahead reaching down the length of two continents was a long, grim gantlet of storm, foe and death.

Yet even now, before the austere flatlands of the arctic had totally disappeared in the horizon mists behind him, the curlew was feeling the first faint stirrings of another year's mating call that would drive him back to await the female when springtime greened the arctic lichens again.

The Gantlet

. . . Being the hitherto unpublished notes of Lucien McShan Turner on the birds of Ungava. . . . I saw no Esquimaux curlew until the morning of the 4th of September, 1884, as we were passing out from the mouth of the Koksoak River. Here an immense flock of several hundred individuals were making their way to the south. . . .

Chapter Three

The curlew's wings beat with a strong, rapid, unchanging rhythm hour after hour. The strokes were deep, smooth and effortless, the wings sweeping low beneath his belly at every downstroke and lifting high over the back with each return. Each stroke was an intricate series of gracefully coordinated actions merged with split-second precision into a single, smooth movement, for the curlew's wing was a wing and propeller combined. Each portion of the wing had a different flight role to play.

The sturdy inner half next to the body deflected the airstream as an aircraft wing does, so that pressure developed against the undersurface and suction above—the lift that produces flight. It accomplished this with its aerodynamic shape alone. The flapping of the wings provided forward drive, but was not directly responsible for keeping the bird airborne.

The outer half of the curlew's wing, composed of the ten stiff flight feathers overlapping like a venetian blind, was the propeller that drove the bird forward, producing the airflow that gave lift to the inner wing. The wing bones were along the leading edge of the wing, and most of the wing behind them consisted of flexible feathers. Hence on the downstroke the pressure of air twisted the wing, bony front edge down and trailing feather edge up, which turned it into a propeller blade pushing air backward and driving the bird ahead. The wingtip flight feathers were additional small propellers, for each central quill was also nearer the front than the rear, so that air pressure twisted each one individually, the same way the whole wing twisted, pushing air backward and adding forward drive. With the upstroke, the air pressure bent the wing and feathers the opposite way, now front edges up and rear edges down, so that the push of feathers against the air still produced a forward propulsion, and the lift force of the inner wing remained uninterrupted with no loss of altitude on the upstroke.

It was all reflexive, automatic, too rapid for conscious control. The curlew completed three or four wingbeats a second to give him a flight speed of fifty miles an hour.

Occasionally one of the curlew's wings would bite into the harder, spiraling air of a vortex left by the wingtips of a migrating shorebird ahead of him, for even the passage of another bird left a trail in the air that the curlew's delicately sensitized wings could detect. Usually this alteration in the air pattern was the curlew's first warning that he was overtaking a flock of birds ahead. When he found one of these vortexes, the curlew took advantage of it and followed it in with one wing riding the updraft edge of the horizontal column of spiraling air. In this way he found a degree of lift ready-made for him and his own wings could work a little easier.

But no other shorebird except the golden plover flew as fast as the curlew did, and each time he slowly overtook the bird producing the vortex ahead. First he would hear the faint twitter of a flock's flight notes, the vortex would grow stronger, then the birds would appear as blurred figures against the grey sky in front. The curlew would fly with them for a time, but his greater speed would gradually drive him ahead. Then once more he would be flying alone.

This happened several times during the night, for the air layers close to the cooling tundra were turbulent and most of the shorebirds were flying at the same level just above the turbulence. Toward morning the curlew encountered another vortex trail and adjusted his wingbeat to the change in lift. He followed it for a long time and the vortex remained firm but grew no stronger. This time the curlew wasn't overtaking the flock ahead. Ducks and geese were not yet migrating, only two birds could be flying out of the arctic now at a speed that the Eskimo curlew wouldn't rapidly overhaul. They had to be either golden plover or his own species, Eskimo curlew.

The curlew's tireless wings beat faster and the airflow pressed hard against his streamlined body. The wingtip vortex eddying back from the unseen flyers ahead strengthened, and it was a firmer, rougher vortex than any the curlew had encountered earlier in the night. It grew stronger almost imperceptibly, and the curlew's eagerness grew with it. A tenuous hope, part instinctive reaction and part a shadowy form of reasoning, formed nebulously in the curlew's brain. Was this the end of his lifelong quest for companions of his own kind? The curlew's wingbeat speeded until the powerful sinews of his breast muscles, gram for gram among the strongest of animal tissue on earth, pained with the strain.

The other birds were very close before their figures emerged, faintly at first and then more sharply, out of the darkness ahead. For a minute or more the curlew could detect only the vague, wavering lines of the flock's formation, then slowly the dark lines separated into individual birds. Only the fast, strong flyers like geese, curlew and golden plover flew in single-column, diagonally trailing lines or Vs that permitted each bird to benefit from a wingtip vortex of the bird ahead yet escape the air turbulence directly behind it. And the curlew knew that the geese flocks were not yet migrating south. A restive excitement seized him and the curlew pushed on harder.

The gap closed rapidly and the birds ahead assumed sharper form. They were small, much smaller than the curlew, yet now there was not the instinctive rejection that had caused him to ignore the Hudsonian curlews and other shorebird flocks. The urge to join a flock was still as pressing as before. The curlew called out softly. Golden plovers answered.

It was a large group of forty or fifty, and the curlew moved in to a rearguard spot at the trailing end of one of the arms of the flock. He slackened flight speed and announced his presence with a rapid, twittering series of notes. The plovers answered again, the whole flock chattering sharply in unison. The curlew's flocking urge was satisfied. There was a vague, remote feeling of loneliness deep within him still, but the curlew was no longer alone.

Of the thirty-odd shorebirds that fly south out of the Canadian arctic every fall, only the golden plover is suited as a migration companion for the Eskimo curlew. Their flight speeds and food preferences are similar, but there is another more important reason. With their tireless endurance as flyers, the golden plover and Eskimo curlew spurn the easy land route down the continent that all other migrating birds follow. Instead they work eastward to the rocky coasts of Labrador, Newfoundland or Nova Scotia, then strike out straight south over the Atlantic for a gruelling, nonstop flight of twenty-five hundred or more miles that doesn't bring them to land again until they reach the northern shores of South America forty-eight hours later. Often a big Hudsonian godwit or, occasionally, a shorebird of some other species will join a golden plover flock and follow the plovers down the Atlantic on this long oversea shortcut south. But only the Eskimo curlew and golden plover do it regularly every fall, for only they, of all the arctic's strong-winged shorebirds and waterfowl, possess the speed and power of flight to breast or escape the midocean storms often encountered. The route enables them to take advantage of the rich crowberry crop that purples the hillsides and plateaus of the Labrador peninsula each fall, a luxuriant store of food missed by the hosts of midcontinental migrants. But in spring the plovers and curlew must follow the usual migratory route up the western plains. For then the crowberries are dead and hard beneath snows of the Labrador winter that linger for weeks after the midcontinent's arctic is greening with spring.

Toward dawn the grey monotony of tundra, dimly visible far below, began to be pierced by slender, twisting fingers of black. The birds had covered four hundred miles since nightfall and were approaching the tree line where tundra gave way to the matted subarctic forests of spruce. The black fingers reaching into the tundra were forested river valleys where stunted spruce thickets found shelter in the hollows against winter blizzards and precariously survived. With the first yellow-grey flush of dawn the flock dropped to a lakeshore mudflat, rested briefly, then as daylight came they began busily feeding.

The curlew with his stiltlike legs and long, downcurved bill stood out strikingly among the smaller, dark-plumaged, short-billed plovers. But the two birds, competitors and enemies on the nesting grounds, had migrated in company for countless generations and they mingled now as one species. Other shorebirds—yellowlegs, knots and the little semipalmated and least sandpipers—scurried close in their feeding, then withdrew. The curlew studied them closely, for somewhere in this vast arctic tundra were birds he would recognize as his own kindred.

They fed all day with only occasional breaks for resting. With the darkness they flew again. The flock clung together loosely as they climbed for height, then as they leveled off the birds formed smoothly into a straggling V formation that permitted the inner wing of each bird to gain support from the whirling air produced by the outer wing of the bird ahead. The curlew took the lead position at the point of the V and the plovers fell in behind with a grace and ease as though the maneuver had been long practiced. No conscious selection of flock leader had taken place. The bird at the point position had to work harder to create lift and forward speed out of the unbroken air barrier ahead of it, and the curlew was the strongest flyer, so the remainder of the flock formed automatically behind in a movement as involuntary and spontaneous as each bird's breathing.

Soon after starting, the black fingers below merged into a solid mat. They were over spruce forest now and the tundra was behind. Other shorebirds were flying straight south toward the western plains, but the curlew led his flock southeastward, veering toward the matted crowberry vines of Labrador. Occasionally the curlew dropped back to an easier flight spot in the body of the flock, but each time after a brief rest he moved forward to the lead again.

The Gantlet

PROCEEDINGS *of the*

ACADEMY OF NATURAL SCIENCES

of Philadelphia, 1861

August 13th. Dr. Leidy in the chair. Nine members present. The following papers were presented for publication: "On Three New Forms of Rattlesnakes," by Robert Kennicott. "Notes on the Ornithology of Labrador," by Elliott Coues. . . .

The Esquimaux curlew arrived on the Labrador coast from its more northern breeding grounds in immense numbers, flying very swiftly in flocks of great extent, sometimes many thousands. . . . The pertinacity with which they cling to certain feeding grounds, even when much molested, I saw strikingly illustrated on one occasion. The tide was rising and about to flood a muddy flat of perhaps an acre in extent, where their favorite snails were in great quantities. Although six or eight gunners were stationed on the spot and kept up a continual round of firing upon the poor birds, they continued to fly distractedly about over our heads, notwithstanding the numbers that every moment fell. They seemed in terror lest they should lose their accustomed fare of snails that day. . . .

By order of the LIBRARY AND PUBLISHING COMMITTEE, *the following proceedings of the* BOSTON SOCIETY OF NATURAL HISTORY *for 1906–7 are published . . .*

Paper No. 7—Birds of Labrador. By Charles W. Townsend, M.D., and Glover M. Allen . . . Numenius Borealis *(Forster), Eskimo curlew. Formerly an abundant but now a very rare autumn transient visitor in Labrador.*

> *When August comes if on the Coast you be,*
> *Thousands of fine Curlews, you'll daily see.*

Packard writes of the curlew as follows: "On the 10th of August, 1860, the curlews appeared in great numbers. We saw one flock which may have been a mile long and nearly as broad; there must have been in that flock four or five thousand. The sum total of their notes sounded at times like the wind whistling through the ropes of a thousand-ton vessel. . . ."

But we met with none during our visit to the Labrador coast in the summer of 1906. We talked with many residents and they all agreed that the curlew though formerly very abundant, suddenly fell off in numbers, so that now only two or three or none at all might be seen in a season. Capt. Parsons of the mailboat Virginia Lake *said that they were very abundant up to thirty years ago. He often shot a hundred before breakfast, often killing twenty at a single discharge. Fishermen killed them by the thousands. . . . They kept loaded guns at their fish stages and shot into the flying masses, often bringing down twenty or twenty-five at a discharge.*

To sum up the evidence, we can state that the natives of Labrador persistently harassed the Eskimo curlew but did not realize there was any diminution in their numbers until about 1888 to 1890. After 1892, but a small remnant of this formerly abundant bird has visited Labrador's shores. . . . It is apparent that they are now a vanishing race—on the way to extinction.

Chapter Four

Nights of endless flying and days of feeding at the edges of stagnant muskeg ponds followed monotonously. The green flashes of the arctic sky's aurora borealis grew fainter behind them and they reached each dawn with hardening breast muscles that felt no fatigue. On the James Bay salt marshes food was abundant and they lingered for many days, gorging on the minute water and mud life until the southland call pressed them on again. The curlew led them straight eastward now over the ancient eroded mountains of Quebec toward the gnarled gneiss seacliffs of Labrador's Gulf of St. Lawrence.

On the second morning the dawn came in foggy and cold. There was a sharp salty tang in the heavy air that struck their nostrils and the curlew led the flock without stopping as the dawn brightened imperceptibly into a grey, sunless day. The air warmed and the fog banks thinned and here and there brown-green patches of the bare, craggy coastal plateau began appearing between the dispersing windrows of mist below them. Then the salt tang sharpened and the fog grew thick again and the curlew knew they were nearing the seacoast. There was nothing ahead, above or below, but the pressing whiteness of fog, but the curlew held course unerringly. Suddenly, toward midmorning, the enveloping whiteness was pierced by the rumble of surf and screeching of gulls. The curlew banked sharply and dropped in a steep dive, zigzagging erratically to control the speed of descent. The plovers broke their flock formation and followed the curlew down. They leveled off a few feet above the water, oriented themselves with the direction of wave movement and followed the wave crests in until the cliffs broke out of the fog in a giant rock wall that towered above them. The curlew had been flying blind for several hours, but he had overshot the coast by less than a mile.

They climbed again, skimmed across the cliff top and landed. Creeping, heathlike vines of the crowberry lay everywhere and in patches the fleshy, purple berries were so thick they hid the foliage. The birds commenced feeding immediately. The wind off the sea was cold and laden with fine rain. After an hour they stopped feeding and bunched together, each bird standing with its head into the gale so that the wind carried the rain back along its overlapping feathers and off its tail.

For two weeks now there would be nothing to do but gorge and fatten for the long, nonstop flight down the Atlantic to South America. It was mid-August and the Labrador summer was already almost gone. The nights were frosty; the days were days of interminable fog. They ate crowberries until their legs and bills and plumage and excrement were stained purple with the juice. On the odd day when the fog lifted under a warming sun they flew to the beaches at low tide periods to gorge on snails and shrimps.

Every day they encountered at least one other flock of golden plovers and the curlew would stop its feeding to scan the passing flocks for another curlew like himself. There were no other curlew, no other shorebirds of any species except the plovers. But other birds were numerous. Gulls, screeching raucously into the fog, were everywhere. Long black and white lines of eider ducks were passing endlessly offshore. The stubby winged and clumsy-flying auks and murres were still gabbling and fighting on the cliff ledges where they had raised their summer broods.

Relatively inactive now, the curlew and plovers fattened quickly. Their breasts were soft and round again with the fat layers that covered the rigid muscles beneath. August had almost ended when the old restlessness seized them again. On days when the weather cleared and the wind was right thousands of other plovers climbed high and left the coastline on a course straight south across the Gulf of St. Lawrence toward the vast Atlantic beyond. But the curlew waited, held by a tenuous bond that his meager brain felt but couldn't quite identify. Vaguely he sensed that when the Eskimo curlews of the tundra came, they would have to come this way.

The restless urge to push on grew stronger and the curlew was torn between the two torturing desires—to wait or to move on. He found partial release from his restlessness by leading the flock on long flights up and down the coastline.

Then the plovers began breaking away, joining in twos and threes with other southward flying plover flocks. The flock had dwindled to half its original size when September came and the nights grew suddenly colder. Now the fog banks that rolled in off the sea occasionally carried big, wet flakes of snow. The last plover flocks had gone. The curlew's flock was alone with the gulls and eiders.

Frost had hardened the crowberries and with their succulent juices gone the feeding had become sparser. The fat that the birds had stored up as body fuel for their ocean flight was beginning to be reabsorbed before the flight had even begun.

Finally the curlew could restrain his migratory urge no longer. On a cold dusk after a blustery day during which the temperature had barely risen above the freezing point the curlew took wing and climbed into the murky sky. The cloud ceiling was low and the flock leveled off quickly and turned seaward into a strong head wind. At this altitude it was a full gale that cut their flight speed in half. Gusts periodically broke up the flock formation. Several weaker plovers dropped behind. The curlew knew before the jagged Labrador coastline was lost to sight behind that they couldn't go on. He turned back and in a few minutes the flock alighted

again on a hillside lee where the gale whined distantly overhead.

Having once started and failed, the curlew and plovers
were eager now to begin the long flight. But there were no
more false starts. The curlew waited restlessly for suitable
weather, but the season was now late and suitable days were
few. The fog dispersed but the south gale blew without break
for three days and three nights while the birds fed intermittently on the drying crowberries and beach snails. On the
fourth day the gale swung around the compass and continued, lighter now but colder, out of the north. This, a tail wind,
was what the curlew awaited. Night came cold and clear.

The takeoff, the climb for height, the automatic V-ing with
the curlew at the point were accomplished with the same ca-
sual unthinking precision as on numerous dusks before. The
curlew and many of the plovers had made the ocean flight in
previous autumns and they had a shadowy, remote memory
of it. Most of them sensed obscurely that when dawn came
there would be only the vacant sea below their wings, that
they would fly on and on and another night and another
dawn would come and the same vacant sea would still be
there. And they knew that the sea was an alien and hostile

element, for they were strictly creatures of the land and of the air. During periods of unusually smooth water they might alight briefly on the ocean's surface to snatch a few moments of rest, but they were clumsy swimmers at best, their feathers lacked oil and waterlogged quickly, and rarely did the sea provide the calm conditions that would permit even a momentary landing. Usually the long flight, once begun, had to be completed nonstop without food for their stomachs or respite for their wings.

Behind them now the arctic's aurora borealis was flashing vividly above the Labrador skyline, but when they came to earth again, with flight feathers frayed and their breast muscles numbed by fatigue, it would be in a dank jungle river bottom of the Guianas or Venezuela. Yet there was no fear or hesitation now with the takeoff, no recognition of the drama of the moment. There was only a vague relief to be off. For it was a blessing of their rudimentary brains that they couldn't see themselves in the stark perspective of reality—minute specks of earthbound flesh challenging an eternity of sea and sky.

The Gantlet

For the increase and diffusion
of knowledge among men.
SMITHSONIAN INSTITUTION, WASHINGTON.
ANNUAL REPORT *of the*
BOARD OF REGENTS
for the year ending June 30, 1915. . . .

In Newfoundland and on the Magdalen Islands in
the Gulf of St. Lawrence, for many years after
the middle of the nineteenth century, the Eskimo
curlews arrived in August and September in millions
that darkened the sky. . . . In a day's shooting by
25 or 30 men as many as 2,000 curlews would be
killed for the Hudson Bay Co.'s store at Cartwright,
Labrador.

Fishermen made a practice of salting down these
birds in barrels. At night when the birds were roost-

*ing in large masses on the high beach a man armed
with a lantern to dazzle and confuse the birds could
approach them in the darkness and kill them in enormous numbers by striking them down with a
stick. . . .*

Chapter Five

The curlew held to a course that was almost due south. When the tumbling Labrador hills dropped from sight behind, the last orienting landmark was lost, but the curlew led the flock unerringly on. Somewhere in the cosmic interplay of forces generated by the earth's rotation and magnetic field was a guide to direction to which hidden facets of his brain were delicately tuned. He held direction effortlessly, without conscious effort. An unthinking instinct, millenniums old, was performing subconsciously a feat beyond the ken of the highest consciousnesses in the animal world.

The night was but yet half spent when white surf outlined the craggy coastline of Nova Scotia's Cape Breton half a mile below. On some other years the curlew had stopped here, but the season was late and there was no thought of stopping now. It had taken five hours to cross the Gulf of St. Lawrence and the flock pushed now without pause across the tip of Cape Breton to the 2,500-mile misty maw of the Atlantic beyond.

The curlew dropped back for rest to an easier flight spot in the body of the flock and stayed there an hour while one of the plovers led. Then a cold front of air, moving eastward off the Canadian mainland, enveloped them in an area of turbulent air currents and the curlew moved forward to the lead again. The warm lower layers of air were being lifted by the heavier cold air pushing beneath. In the colder temperature of higher altitudes, the warm air's moisture began condensing, first into misty rain then, as its temperature dropped, it became snow.

Erratic air currents buffeted the flock and the formation broke up. The snow, light and sparse at first, became thicker. The flakes grew into large, loose, damp clusters that caked into the birds' wing feathers and made flight difficult. The curlew, reacting instinctively, led the flock upward in a steep spiraling climb. The air turbulence decreased as they climbed, but the snow clouds grew denser. The quieter air permitted them to line up in formation again, but they had to form ranks more by the feel of the wingtip air whorls than by sight, for now the snow was so thick that frequently even the bird next ahead was hidden. They stopped climbing and leveled off again.

There was no way of detecting how fast the cold front was moving eastward, but the curlew knew—partly from half-remembered experiences of previous migrations, but mostly by an instinctive intuition—that their fifty-mile-an-hour flight speed would take them back through the front and keep them ahead of it, because the storm's front would be moving at a speed slower than theirs. But they would have to turn and fly with the storm, and that was eastward toward mid-Atlantic.

The curlew veered eastward and the double rank of plovers behind followed his deflecting air trail, though only the front few birds had been able to see the curlew turn. The snow clung to their wings, packed into the air slots between the flight feathers. Wings that a few minutes before had responded deftly to the gentle, rhythmic flexing of the breast muscles were now heavy and stiff, and they beat the air futilely like lifeless paddles, driving air downward in a waste of energy instead of deflecting it rearward for the horizontal airflow essential to flight. Their flight speed dropped until they were hovering almost motionless in a disorganized, bewildered cluster, now almost a mile above the sea. Then the curlew led them eastward again by angling slowly downward and drawing from gravitational pull the flight speed that their soggy wing feathers could no longer produce unaided. Now their flight speed was normal once more, but they were sacrificing altitude rapidly to maintain it. Up from the grey void below, the sea was rising steadily toward them.

The curlew led them on a long gradual, seaward incline, adjusting the downward flight angle to the pressure of the airflow on its sensitized wings so that normal speed was maintained with the minimum of altitude loss that would accomplish it. Occasionally the snow thinned and for brief intervals almost level flight was possible. Then it thickened again and their wings grew heavy and the curlew would have to angle sharply downward.

Behind them, but cut off probably by several flight-hours of impenetrable snow, were the coastlines of Nova Scotia and

New England. Ahead, perhaps only minutes, was the storm front with warmer undisturbed air before it. But even if the storm front were overtaken and passed there was only a limitless Atlantic beyond, into which they would have to keep flying to stay ahead of the snow clouds now pressing them implacably toward the wave crests below. All this the curlew knew, not from any process of reasoning but via the same nebulous channels of instinct that told him too that somewhere a mate of his own species was waiting for another breeding time to green the tundra lichens again.

Even the curlew's thick breast muscles and wing tendons, stronger by far than those of the smaller plovers he led, were aching and burning now from the abnormal energy output demanded to overcome the effect of the crusting snow on their wings. Their downward flight course took them into the lower layers of turbulent, bumpy air once more. The flock was thrown out of formation again. They clung together by calling sharply and constantly to each other. Each bird was alone in a gusty white world of its own, unseen and unseeing, but the quavering chatter of flight notes was a nexus that held them together.

For a long time the blind, numbing flight continued and the curlew fought to maintain height until not only his breast muscles but every fiber of his body throbbed with agonizing fatigue. To the lisping murmur of flight notes from the plovers behind there was soon added a sibilant hissing that came from below. The hissing grew stronger. It was the sound of snow striking water.

Then through the white curtain the curlew could see it. Waves with silvery caps curling upward appeared first ahead of the flock, paused momentarily below as they were overtaken, then disappeared behind. The snow had cleared slightly and now the plovers became visible again strung out haphazardly to the curlew's rear. The hindmost, weaker birds were lower, closer to the sea. They had had to sacrifice altitude faster to keep up with the stronger flyers ahead. Glutinous snow clung to their wingtips, the melting rate from body heat barely equalling the rate at which new snow accumulated.

The curlew would hold a level plane of flight for several seconds, then as forward speed decreased he would have to dip downward, gain new speed and level off again. The sea was clearly visible now, the white wave crests etched sharply against the black water. At times a higher crest leaped upward to within a few feet of the struggling birds.

A great wave appeared ahead. The curlew fought the lethargy in his wings and lifted over it painfully to drop into the trough beyond. He struggled on. The next crest was lower and the curlew mounted it with several feet to spare. Behind him, the great wave lunged into the plover flock. Three of the lower birds fought for height but could do no more than hover helplessly. There was no cry. The wave arched upward momentarily and the birds disappeared from sight. The wave passed and the three plovers didn't reappear.

Nature, highly selective in all things, is most selective with death. The weak neither ask nor obtain mercy.

The flock slogged on, a few feet above the sea, struggling laboriously over each crest and snatching a few niggardly seconds of partial rest in the quieter, protected air of each trough. Once a long trough lifted into a seething comber many feet higher than those preceding and the spray of its crest lashed the curlew's wings. The curlew had to battle a maelstrom of air currents for several seconds to keep airborne. When the wave passed two more of the plovers failed to reappear. But the spray melted much of the snow clinging to the curlew's wingtip feathers. For a minute his unburdened wings could bite into the air with all their old power. Then

the snow clogged them again. Only the knowledge that somewhere close ahead the cold front terminated kept the curlew plunging on.

The air grew warmer very gradually, so slowly it was difficult to detect the change. But the snow altered to rain abruptly. At one wave crest there was only the swirling white wall of snow ahead, by the next crest the snow was behind and sheets of rain pelted them. The snow melted from their feathers in a few seconds and the curlew led the remnant of his flock upward in a sharp climb. The pain and fatigue drained quickly from their wings and breasts with the resumption of normal flight. The sea disappeared again in the darkness beneath them. After several minutes they broke through the rain front into a quiet mist-roofed world beyond.

The Gantlet

. . . *And sometimes, during northeast storms, tremendous numbers of the curlews would be carried in from the Atlantic Ocean to the beaches of New England, where at times they would land in a state of great exhaustion, and they could be chased and easily knocked down with clubs when they attempted to fly. Often they alighted on Nantucket in such numbers that the shot supply of the island would become exhausted and the slaughter would have to stop until more shot could be secured from the mainland.*

The gunner's name for them was "dough-bird," for it was so fat when it reached us in the fall that its breast would often burst open when it fell to the ground, and the thick layer of fat was so soft that it felt like a ball of dough. It is no wonder that it was so popular as a game bird, for it must have made a delicious morsel for the table. It was so tame and unsuspicious and it flew in such dense flocks that it was easily killed in large num-

bers. . . . Two Massachusetts market gunners sold $300 worth from one flight. . . . Boys offered the birds for sale at 6 cents apiece. . . . In 1882 two hunters on Nantucket shot 87 Eskimo curlew in one morning. . . . By 1894 there was only one dough-bird offered for sale on the Boston market.

Chapter Six

The curlew knew that they had to continue flying eastward to keep the storm from overtaking them again, but it was a simple, uncolored, matter-of-fact knowledge. There was no lingering emotional reaction, no fear. The terrors of the snow-filled sky, the plovers forced into the sea, were forgotten almost immediately. Only the fact of the storm itself was remembered, and it was remembered not in panic or fright, but merely as a natural foe that was there and had to be avoided.

But their course eventually had to be southward, not eastward. To the east for four thousand miles there was only empty sea. After half an hour the curlew turned the flock southward, and they flew south unhindered for almost another half hour before the eastward-moving storm front enveloped them again. At the first big drops of rain, the curlew veered sharply to the east once more and in a few minutes the flock reentered clear air.

In the three hours that remained before dawn, they repeated this many times, flying south until the rain overtook them, then veering eastward to get ahead of it again. They were on a southerly course when the yellowing dawn pierced a murky eastern sky. Daylight came swiftly, changing the black of the sea to a cold green, but there was no sun. They flew southward for an hour, then two hours, and the cloud cover grew thinner and the day brightened and this time the storm didn't reappear. Even the grey, bumpy clouds of the western sky vanished and in the east the sun cut like a torch through the dissolving mists. The air remained cold, but in a short time the sun stood alone in a blue and otherwise empty sky.

The birds had worked southward around the storm. The snow clouds of the night, what would be left of them, would be breaking up now far to the north over the codfish shoals of Newfoundland's Grand Banks.

In midmorning the air warmed and eddying wisps of fog began rising off the sea. The sky above remained blue and clear, but at times the sea below was completely hidden by layers of mist. They were approaching the spot where the icy Labrador current flowing southward out of the arctic met the tepid northward-flowing tropic waters of the Gulf Stream. Here the Gulf Stream is deflected eastward past Newfoundland into mid-Atlantic. After an hour of intermittent fog the sea lay bare again. Then its pale green arctic waters changed abruptly to a deep indigo blue with a line of demarcation as sharp as a line between water and shore. They were over the Gulf Stream, a product of the tropics. The green of the Labrador current, last feature of the arctic, faded behind them.

Their wings beat mechanically, without change of pace or fatigue. The air warmed constantly, for each hour put them fifty miles southward. The only change in the drowsy monotony of flight came when, at intervals, they let themselves drop low to skim the wavetops for perhaps an hour before climbing again.

At low altitude the sea, like the arctic tundra, revealed that its surface mask of lifeless barrenness was illusion. Life was there, abundantly, when the birds came low enough to see it. At times shimmering discs of jellyfish dotted the sea for miles; the sun glinted metallically off a thousand silver flanks as schools of small fish darted upward into the surface layers; and sometimes there were clouds of minute one-celled plankton creatures, each one a microscopic grain of orange pigment by itself, but in their billions they colored miles of sea a vivid red.

Down close to the water, there were other birds too, birds that spent most of their lives skimming the open vistas of sea, touching land only when the irresistible urge of the nesting time drove them ashore. Wilson's petrels fluttered mothlike and dodged erratically between wave crests, their white rumps flashing like tiny breakers as they fed incessantly on the sea's crustacea and plankton. Phalaropes that had nested on the arctic tundra with their shorebird kin had returned now to the sea, which would lure them until another nesting time came. Occasionally a bigger shearwater soared past on black, motionless wings that skilfully utilized the updrafts created atop each wave crest by the upward deflection of surface wind. But these were true birds of the sea. The sea gave them food, and when their wings tired the sea also gave them rest, for they swam as skilfully as they flew.

The curlew and plovers could only keep flying, waiving food and rest until the landfall came.

By evening they had crossed the eastward-flowing arm of the Gulf Stream and were over the immense two-million-square-mile eddy of the mid-Atlantic where no currents came to stir the brackish water and where the rubbery fronds of sargassum weed collected in the great floating islands of the Sargasso, weirdest of all seas. They had flown almost twenty-four hours, yet there was no fatigue in the pulsing muscles of their breasts.

Vast meadows of brown floating algae passed beneath. At intervals when the birds came low, they would see flying fish with great winglike pectoral fins extended, skimming over the soggy knots of seaweed. There were crabs, shrimps and sea snails clinging to the seaweed stems. In other years this first dusk had put the curlew within sight of Bermuda's flat-topped Sear's Hill, but the night's storm had driven them far to the eastward, and now the sun set in an empty sea. When darkness came, the sea flamed with the cold white light of millions of phosphorescent creatures.

The curlew led the flock upward and throughout the night they flew steadily at a height of a half mile or so, the birds calling intermittently to each other. When the curlew was leading the flock his senses had to be kept sharply tuned to the vagaries of wind and the cosmic impulses that his brain interpreted into a sense of direction. When he dropped back for rest, he could fly in a half-sleep, his wings beating automatically, his eyes half shut, following subconsciously the trailing air vortex of the bird ahead of him.

That night the North Star and the familiar constellations of the arctic sky dropped almost to the northern horizon. New star groups rose to the south. And shortly before dawn the wind freshened, a warm, firm wind that blew with monotonous constancy out of the northeast. They had entered the region of the trade winds. It was a quartering tail wind that gave them almost another ten miles an hour of speed.

Day, when it came, was hot despite the wind. Occasionally the grey-blue form of a shark glided close to the sea's surface. This was the rim of the tropics, and the sea turned bluer, and condensation of the hot rising air gave the sky a lumpy patchwork of white cumulus clouds. The cloud shadows dappled the blue water with constantly changing patterns of grey. Occasionally there were thicker knobs of cloud that hung motionless on the western horizon, the island signposts of the sea, for every island had its cap of cloud that was visible far beyond the island's own horizons. These were the

Lesser Antilles of the outer Caribbean. And far beyond the rim of the sea, ahead, another twelve hours of flying away, were the jungles and mountains of South America.

Now their breasts and wing tendons were tiring from the thirty-six hours of flying behind them. Flight was no longer the effortless subconscious reflex it had been. It had become a function that had to be willed, only conscious concentration on the task kept their flagging wings working. Two nights and a day without food had slowed their body processes. Now they had to pant rapidly in the hot tropic air, their bills slightly agape, to capture the oxygen supply their lungs demanded. Three of the plovers, one-year-olds making the long ocean flight for the first time, dropped slowly behind and the curlew at the point of the flock slowed to a flight speed that the weaker birds could maintain.

The curlew knew that where the thick clouds dotted the western horizon there were islands only an hour or two's flight away. But he possessed an instinctive knowledge, developed through millennia of his species' evolution, that there was not enough food on small tropical islands to feed the numbers of shorebirds making this ocean flight, and hence he held instinctively to the original course. And he knew that long before the South American coast could be reached a third night would be upon them. Then the landfall would come in darkness and if the night were cloudy and black there could be no landing even then until the dawn light revealed the outlines of Venezuela's mangrove swamps and river sandbars.

The day passed with interminable slowness, the sun sank finally into the Caribbean and the night dropped quickly without twilight. Then the overcast moved in to shut out moon and stars, and rain began falling, for they were reaching the tropics at the height of the rain season. But it was a light, fine rain that cooled the air and made breathing easier. And it was a signal that the coast was approaching.

For another two hours they flew through rain. The curlew could see nothing, but he knew immediately when they left the sea and were flying over land. First the rumble of surf came up through the darkness, then the air became turbulent with the thermal updrafts lifting off the warmer land.

They could do nothing but fly on for hours longer. And now, with the knowledge that land lay below, the continuance of flight became the harshest ordeal of all. Every wingbeat was a torturing battle with lethargy and fatigue. And much of the energy used was now wasted, for their flight feathers were frayed and ragged, no longer capable of the sharp, propelling bite of feather against air, which had made flights so easy and effortless when they left Labrador.

The curlew knew that once they had crossed the coastal strip with its beaches and river estuaries, there was nothing beyond for a hundred and fifty miles but the dense tangle of mangrove swamp where a landing was as impossible as on the open sea. Now, even if the night cleared, they would have to push on regardless until the flat, grassy llanos of the Venezuelan interior spread out below them. Despite the growing heaviness of their wings, the curlew led them upward to clear the coastal mountains he knew were ahead. The climb was a torturing anguish. They leveled off, but it brought no respite to the burning pangs of fatigue that throbbed in every fiber of their small bodies.

The night remained black. At last the dawn came, not yellow or red, but in a somber pall of greyness. The land below was a drowned and sodden land of mud, water and swollen rivers, like the springtime tundra of the arctic. The broad treeless valley of the great Orinoco spread in every direction as far as the grey pall would let them see. The rain still fell.

They had flown without rest or food for almost sixty hours. From a land of snow and the northern lights, they had come nonstop to a land that was steaming with the rank growth of the tropics. Below them were hundreds of miles of mudflats and grassy prairie that teemed with the abundance of aquatic insect food that only the months of tropical rain could produce.

With the first misty light of the dawn, the curlew arched his stiffened wings and plunged downward in an almost vertical dive. He had spanned the length of a continent since his wings had last been still. The plovers followed. The flock touched down.

But not a bird rested, for feeding had to come first. Their stomachs had been empty fifty-five hours and they had flown close to three thousand miles on the fuel stored in Labrador as body fat. When the flight began, their breasts had been round and swollen. Now they were gaunt, constricted, the breastbones protruding in sharp ridges through the feathers. In less than three days each bird had lost more than a third of its body weight—two ounces for the plovers, four or five for the curlew.

They fed rapidly until midmorning, and only then did they rest. On the broad savannahs abutting the Orinoco, food was abundant. They fed again for several hours before the first tropic night brought darkness.

The Gantlet

This is the eighth in a
series of bulletins of the
UNITED STATES NATIONAL MUSEUM
on the LIFE HISTORIES
OF NORTH AMERICAN BIRDS
by ARTHUR CLEVELAND BENT.

Order Limicolae. Family Scolopacidae. . . . Numenius borealis, *Eskimo curlew. . . .*

Excessive shooting on its migrations and in its winter home in South America was doubtless one of the chief causes of its destruction. . . . I cannot believe that it was overtaken by any great catastrophe at sea which could annihilate it; it was strong of wing and could escape from or avoid severe storms; and its migration period was so extended that no one storm could wipe it out. There is no evidence of disease or failure of food supply. No, there was only one cause, slaughter by human beings, slaughter in Labrador and New England in summer and fall, slaughter in South America in winter, and slaughter, worst of all, from Texas to Canada in the spring. They were so confiding, so full of sympathy for their fallen companions, that in closely packed ranks they fell, easy victims of the carnage. The gentle birds ran the gantlet all along the line and no one lifted a finger to protect them until it was too late. . . .

Chapter Seven

The plovers and curlew lingered on the savannahs of the Orinoco for two weeks, rapidly growing fat again. There were thousands of other shorebirds flocking the great grasslands—golden plovers that had come down the long oceanic migration route as the curlew's flock had done, and a dozen other species that had followed the land route of the central plains and the Panama isthmus to rendezvous here on the Venezuelan prairies. There were brilliant tropical birds too, now in the middle of their nesting time and busily feeding young. White egrets had covered acres of riverside swamp with their big nests, the nests often so numerous that they touched one another. Flocks of scarlet ibis, the gems of tropical bird life, followed the riverbanks in their food hunts, approaching first as silhouettes of colorless grey, then flaming into a vivid scarlet as they came nearer, and fading to grey again when they passed.

Food was limitless on the llanos, and many of the arctic shorebirds would migrate no farther, but after two weeks of feeding had fattened them once more the curlew and plover felt the old restless torment calling them again to a more distant southland. The other plover flocks had already gone. As in Labrador, the curlew's flock was the last to depart.

They took off on a bright moonlit night early in October, followed a tributary valley of the Orinoco until it lost itself in the mountain range that separated the Orinoco and Amazon watersheds, then dropped into a deep valley of one of the Amazon's tributaries beyond. They followed the slender thread of water southward, and by dawn they had reached the broad Amazon itself. The next night, to take advantage of an eastward shift in the wind, they turned southwest, and another five hundred miles of flying put them, at dawn, within sight of the Peruvian Andes' snowcapped peaks. The wind died, and now for three nights following they flew southeast along the Andean flanks. On the fifth dawn, gaunt and wing-worn again, they dropped to the grassy flatlands of the Argentine pampas, twenty-five hundred miles south of the Venezuelan llanos.

Spring was greening the pampas grass and giant thistle. Grasshoppers were emerging. For days the birds did little but gorge on the insect life of the short grass plains, flying at intervals to the lower levels where the grass grew denser in brackish marshes and swarms of aquatic insects provided a change of diet. They were always moving, but never moving far. Their worn wing feathers were molted one by one and replaced, giving them full flight power again. Here, they were eight thousand miles from the arctic nesting grounds and of all the tundra shorebird species only the yellowlegs, knot, buff-breasted sandpiper and one or two others had migrated so far, yet at times the restless migration urge still pressed the curlew and plovers southward. On clear nights when the prevailing westerlies swept strongly across the prairies, giving them a good beam wind, the flock would take off again. Hours later, another one or two hundred miles southward, the restlessness would be temporarily appeased and the curlew would lead them down to a moonlit knoll to await the dawn.

In this manner they straggled slowly southward. By the time the hot December sun had burned the giant thistles, and the pampas grass was silver with its nodding panicles of flowers, they were deep down into the stony undulating plains of Patagonia, within a single night's flight of the Antarctic Sea. The herculean thrust of the migratory impulse had carried them from the very northernmost to the southernmost reaches of the mainland of the Americas. Yet even here there were still great flocks of shorebirds. The days were long and hot, the brief night cool. Of all the world's living creatures, none but the similarly far-flying arctic tern sees as much sunlight as the shorebirds that spend each year chasing, almost pole to pole, the lands of the midnight sun.

For almost five months the curlew and plovers had been goaded by an insatiable drive that had relaxed at times but never fully disappeared. Now the urge of the migration time was dead. A peculiar lethargy gripped the plovers and they were content to fly back and forth between two salt lagoons —feeding, dozing, flying listlessly, waiting like an actor who has forgotten his lines for the prompting of instinct to tell them what to do next.

But within the curlew, as fast as the pressure of the migratory urge relaxed a new tormenting pressure replaced it. It was the old vague hunger and loneliness. Suddenly the curlew remembered again that he lived alone in a world to which other members of his own species never came. A restlessness of a different sort beset him. He tried to lead the plovers farther afield but they would not follow. Finally the restlessness became irresistible. The curlew spiraled high, circled and recircled the lagoon where the plovers were feeding. He called loudly and repeatedly, but the plovers gave no sign of hearing. Then the curlew turned eastward toward the coastal tide flats that he knew were there, many hours of flight away. He was flying alone again.

Patagonia had none of the deep rich soil of the pampas. Much of it was gravelly shingle, cut by sharp ridges of volcanic rock, and the vegetation was scanty. Even where the coarse grass and thistles grew, they were burned brown now by the fierce summer sun. Out of this arid land the shorebirds were drifting eastward toward the cool, food-rich mudflats of the seacoast.

Here one of the highest tides of the world leaves miles of beach exposed at every ebb, and the stranded flotsam of the sea replenished twice daily was a food supply that never waned. Vast flocks followed each low tide outward. Most of them were golden plover, but there were yellowlegs too, flashing their white rumps, while buff-breasts and sander-lings daintily dodged the breakers as though afraid to get their feet wet.

The curlew wandered from flock to flock, seeking rest-lessly he was not sure what. His long, downcurved bill and wide spread of wing made him stand out prominently among the thousands of other smaller shorebirds.

It was January, and the tundra nine thousand miles to the north would remain for months yet a sleeping, lifeless land of blizzard and unending night, but the curlew began to feel the arctic's first faint call. It was a feeble stirring deep within, a signal that dormant sex glands were awakening again to an-other year's breeding cycle. It was barely perceptible at first. It strengthened slowly. And it was a sensation vastly different from the autumn migratory urge. The call to migrate south had been a vague, restless yearning for movement in which the goal was only dimly defined, but in this new call the goal was everything and the migration itself would be incidental. The essence of what the curlew felt now was a nostalgic yearning for home. And the goal was explicit—not merely the arctic, not the tundra, but that same tiny ridge of cobble-stone by the S twist of the river where the female would come and the nest would be.

The curlew started home. Drifting slowly from mudflat to mudflat, he didn't move far each day, but the aimlessness was gone. The movement was always northward.

The other shorebirds had felt it too. They were constantly moving and the bird population of the mudflats changed with every hour. In a week the curlew was two hundred miles northward.

The Gantlet

The object of the
GENERAL APPENDIX
to the ANNUAL REPORT
of the SMITHSONIAN INSTITUTION
is to furnish brief accounts of
SCIENTIFIC DISCOVERY *in particular*
directions; reports of INVESTIGATIONS
made by collaborators
of the INSTITUTION . . .

The Eskimo Curlew and Its Disappearance

(Reprinted in this annual report after revision by the
author, Myron H. Swenk, from The Proceedings of
the Nebraska Ornithologists' Union, *Feb. 27, 1915.)*

It is now the consensus of opinion of all informed
ornithologists that the Eskimo curlew (Numenius
borealis) *is at the verge of extinction, and by many the*
belief is entertained that the few scattered birds which

may still exist will never enable the species to recoup its numbers, but that it is even now practically a bird of the past. And, judging from all analogous cases, it must be confessed that this hopeless belief would seem to be justified, and the history of the Eskimo curlew, like that of the passenger pigeon, may simply be another of those ornithological tragedies enacted during the last half of the nineteenth century, when because of a wholly unreasonable and uncontrolled slaughter of our North American bird life several species passed from an abundance manifested by flocks of enormous size to a state of practical or complete annihilation. . . .

The Committee on Bird Protection desires to present herewith to the Fifty-fifth Stated Meeting of the American Ornithologists' Union the results of its inquiries during 1939 into the current causes of depletion or maintenance of our bird life. . . . But the most dangerously situated are unquestionably the California condor, Eskimo curlew and ivory-billed woodpecker. They have been reduced to the point where numbers may be so low that individuals remain separated thus interfering critically with reproduction. . . .

Chapter Eight

The arrival of the female was a strangely drab and undramatic climax to a lifetime of waiting. One second the curlew was feeding busily at the edge of the breakers, surrounded by dozens of plovers, yet alone; the next second the female curlew was there, not three feet away, so close that when she held her wings extended in the moment after landing even the individual feathers were sharply distinguishable. She had come in with a new flock of nine plovers. They had dropped down silently, unnoticed except by the sentinel plover that stood hawk watch while the others fed. She lowered her wings slowly and deliberately, a movement much more graceful than the alighting pattern of the plovers. Her long, downward sweeping bill turned toward him.

The female bobbed up and down jerkily on her long greenish legs and a low, muffled *quirking* came from deep within her throat. The male bobbed and answered softly.

There was little mental reasoning involved in the process of recognition. It was instantaneous and intuitive. The male knew that he had been mistaken many times before. He knew that the puzzlingly similar Hudsonian curlews were birds of the arctic summer that he had never seen here on the far-south wintering grounds. He knew this new curlew was smaller and slightly browner, like himself, than the others had been. But these thoughts were fleeting, barely formed. It was a combination of voice, posture, the movements of the other bird, and not her appearance, which signaled instantly that the mate had come.

He had never seen a member of his own species before. Probably the female had not either. Both had searched two continents without consciously knowing what to look for. Yet when chance at last threw them together, the instinct of generations past when the Eskimo curlew was one of the

Americas' most abundant birds made the recognition sure and immediate.

For a minute they stood almost motionless, eyeing each other, bobbing occasionally. The male seethed with the sudden release of a mating urge that had waxed and waned without fulfillment for a lifetime. A small sea snail crept through a shallow film of tidewater at his feet and the curlew snapped it up quickly, crushing the shell with his bill. But he didn't eat it himself. With his neck extended, throat feathers jutting out jaggedly and legs stiff, the male strutted in an awkward sideways movement to the female's side and handed her the snail with his bill. The female hunched forward, her wings partly extended and quivering vigorously. She took the snail, swallowing it quickly.

In this simple demonstration of courtship feeding, the male had offered himself as a mate and been accepted. The lovemaking had begun. There had been no outward show of excitement, no glad display, simply a snail proffered and accepted, and the mating was sealed.

Now they resumed feeding individually, ignoring each other, but never straying far apart. And the cobble bar by the S twist of the distant tundra river called the male as never before.

At dusk he took wing and circled over the female, whistling to her softly. She sprang into the air beside him and together they flew inland over the coastal hills. They landed on a grassy hillside when darkness fell and they slept close together, their necks almost touching. The male felt as if he had been reborn and was starting another life.

They returned to the beaches at dawn and began to move northward more rapidly, alternating flights of ten miles or so at a time with stops for feeding. The call of the tundra grew more powerful and each day they moved faster than the day

before, flying more and eating less. By early February they were a thousand miles north of where they had started, still following the seacoast tideflats, and the annual turgescence of the sex glands with their outpouring of hormones began filling them with a growing excitement. Now the male would frequently stop suddenly while feeding and strut like a game cock before the female with his throat puffed out and tail feathers expanded into a great fan over his back. The female would respond to the lovemaking by crouching, her wings aquiver, and beg for food like a young bird. Then the male would offer her a food tidbit and their bills would touch and the love display suddenly end.

One dusk when the westerly wind was strong off the coastal highlands, they flew inland as they had done every evening, but this time the male led her high above the browning pampas and darkness came and they continued flying. The short daytime flights were not carrying them northward fast enough to appease the growing migratory urge. They left the seacoast far behind and headed inland northwesterly toward the distant peaks of the Andes. Now the male felt a sudden release of the tension within him, for with the first night flight there was recognition that the migration had really begun.

They flew six hours and their wings were tired. It was still dark when they landed, to rest till the dawn. Now they moved little during the day, but at sunset the curlew led his mate high into the air and turned northwestward again. Each night their wings strengthened and in a week they were flying from dusk to dawn without alighting.

They flew close together, the male always leading, the female a foot or two behind and slightly aside riding the air vortex of one of his wingtips. They talked constantly in the darkness, soft lisping notes that rose faintly above the whistle of air past their wings, and the male began to forget that he had ever known the torture of being alone. They encountered numerous plovers but their own companionship was so complete and satisfying that they made no attempt to join and stay with a larger flock. Usually they flew alone.

The northward route through South America was different from the southward flight. When they left the belt of the prevailing westerlies and passed over the pampas into the forested region of northern Argentina, feeding places became more difficult to find. Five hundred miles to the west were the beaches of the Pacific but the towering cordillera of the Andes lay between. From here they could fly northeastward into the endless equatorial jungles of Brazil, where food and even landing places would be scarce for fifteen hundred miles, or they could swing westward to challenge the high, thin, stormy air of the Andes, which had the coastal beaches of the Pacific just beyond. The curlew instinctively turned westward.

For a whole night they flew into foothills that sloped upward interminably, climbing steadily hour after hour until their wings throbbed with the fatigue. And at dawn, when they landed on a thickly grassed plateau, the rolling land ahead still sloped upward endlessly as far as sight could reach, to disappear eventually in a saw-toothed horizon where white clouds and snow peaks merged indistinguishably.

When the sun set, silhouetting the Andean peaks against a golden sky, the curlews flew again. Flight was slow and labored for the angle of climb grew constantly steeper. The air grew thin, providing less support for their wings and less oxygen for their rapidly working lungs. They were birds of the sea level regions and they didn't possess the huge lungs that made life possible here three miles above the sea for the shaggy-haired llamas and their Indian herders. The curlews tired quickly and hours before dawn they dropped exhausted to a steep rocky slope where a thin covering of moss and lichen clung precariously. For the remainder of the night they stood close together resting, braced against the cold, gusty winds.

Daylight illuminated a harsh barren world, a vertical landscape of grey rock across which wisps of foggy cloud scudded like white wings of the unending wind. And the top of this world was still far above them. The peaks that they yet had to cross were hidden in a dense ceiling of boiling cloud.

Nowhere else in the world outside the Himalayas of India did mountain peaks rear upward so high.

Even here, though, there were insects and the curlews fed. It was slow and difficult feeding, not because food was scanty, but because every movement was a tiring effort, using up oxygen that the blood regained slowly and painfully. At dusk the air cooled suddenly and the fog scud changed to snow. They didn't fly. The turbulent air currents and the great barrier of rock and glacier ahead demanded daylight for the crossing.

There was no sleep, even little rest, that night. The wind screeched up the mountain face, driving hard particles of snow before it, until at times the birds could hardly stand against it. Then a heavy blast lifted them off their feet and catapulted them twisting and helpless into dark and eerie space. The male fought against it, regained flight control and landed again. But the female was gone.

He called frantically above the whine of the storm, but his calls were flung back unanswered by the wind. When the wind eased, he rose into the air and flew in tight, low circles, searching and calling, in vain. The wind rose, became too strong for flight, and he clung to the moss of the steep rock

face and waited breathlessly. When it died momentarily, he flew again, but his endurance waned quickly and he couldn't go on. He found a hollow where he could be sheltered from the storm and crouched in it, panting with open bill for the oxygen his body craved. When strength returned he flew out into the wild dark night another time, circling, calling, the agony of loneliness torturing him again.

In an hour he found her, crouched in the drifting snow be-

neath a shelf of shale, as breathless and distraught as he was. They clung together neck to neck and the heat of their bodies melted a small oval in the hard granular snow.

The wind slackened at dawn and the male knew they had to fly, for there could be no lingering here. When the snow changed to fog again and the sun pierced it feebly in a faint yellow glow, they took off and spiraled upward into the flat cloud layer that hid the peaks above. In a minute they were entombed in a ghostly world of white mist that pressed in damp and heavy upon them. They spiraled tightly, climbing straight upward into air so thin that their wings seemed to be beating in a vacuum and their lungs when filled still strained for breath.

In the cloud layer the air was turbulent. Occasionally there were pockets where the air was hard, and their wings bit into it firmly and they climbed rapidly, then the air would thin out again, and for several minutes they would barely hold their own. Once the light brightened and the curlew knew they were close to the clear air above, but before they could struggle free of the cloud a sudden downdraft caught them, they plunged downward uncontrollably and lost in a few seconds the altitude that had taken many minutes to gain.

They broke free of the swirling cloud mass finally and came out into a calm, clear sky. It was a weird, bizarre world of intense cold and dazzling light that seemed disconnected from all things of earth. The cloud layer just below them stretched from horizon to horizon in a great white rolling plain that looked firm enough to alight upon. The sun glared off it with the brilliance of a mirror. A mile away a mountain peak lifted its cap of perpetual snow through the cloud, its rock-ribbed summit not far above. In the distance were other peaks rising like rocky islands out of a white sea.

The curlews leveled off close to the cloud layer and flew

toward the peak. Flight was painful and slow. They flew with bills open, gasping the thin air. Their bodies ached.

As they approached the mountaintop, the wind freshened again. Stinging blasts of snow swirled off the peak into their path of flight. They struggled through and landed for rest on a turret of grey rock swept bare of snow by the wind. Now a new torment racked their aching bodies, for the dry, rarefied air had quickly exhausted body moisture, and their hot throats burned with thirst.

Fifty miles away there were orchids and cacti blooming vividly in the late South American summer, but here on the rooftop of the Americas four miles above the level of the sea was winter that never ended. Not far below their resting place was an eerie zone of billowing white in which it was difficult to distinguish where the snow of the mountainside ended and the clouds began. Yet even here where no living thing could long endure, life had left its mark, for the very rock of the mountain itself was composed largely of the fossilized skeletons of sea animals that had lived millions of years ago, in a lost eon when continents were unborn and even mountain peaks were the ooze of the ocean floors.

The pain drained from their bodies and the curlews flew

westward again past the wind-sculptured snow ridges and out into the strangely unattached and empty world of dazzling sunlight and cloud beyond. They flew a long time, afraid to drop down through the cloud again until there was some clue as to what lay below it, and far behind them the peak grew indistinct and fuzzy beneath its halo of mist and snow. The cloud layer over which they flew loosened, its smooth, firm top breaking up into a tumbling series of deep valleys and high white hills. The valleys deepened, then one of them dropped precipitously without a bottom so that it wasn't a valley but a hole that went completely through the cloud. Through the hole, the birds could see a sandy, desert-like plateau strewn with green cacti clumps and brown ridges of sandstone. It was two to three miles below them, for the Andes' western face drops steeply to the Pacific.

They had been silent all day, for the high altitude flight took all the energy their bodies could produce, but now the male called excitedly as he led the female sharply downward between the walls of cloud. The narrow hole far below grew larger. The air whistled past them and they zigzagged erratically to check the speed of the descent. At first the air was too thin to give their wings much braking power and they plunged earthward with little control, then the air grew firmer, it pressed hard against their wing feathers and they dropped more slowly. Their ears pained with the change in pressure and when they came out below the cloud layer they leveled off again and headed toward the faint blue line of the Pacific visible at the horizon.

Their brief two or three minutes of descent had brought them with dramatic suddenness into a region greatly different from the cold, brilliant void they had left. They were still so high that features below were indistinct, but they were nevertheless a definite part of the earth again. Now there was land and rock and vegetation below them, not an ethereal nothingness of cloud. Here the day was dull and sunless, not glaring with light, but the air was warm. And the air now had a substance that could be felt. It gave power and lift to their wings again and it filled their lungs without leaving an aching breathless torment when exhaled.

They flew swiftly now, for the land sloped steeply and their plane of flight followed the contour of the land downward. Late that afternoon they alighted on a narrow beach of the Pacific. They drank hurriedly of the salt water for a couple of minutes. Then they fed steadily until the dusk.

With twilight the sky cleared and the great volcanic cones of the Andes, now etched sharply against the greying east, assumed a frightening massiveness. Every year the male cur-lew's migratory instinct had led him across this towering bar-rier of limestone, storm and snow. And every year before the memory of it dimmed, the curlew looked back and even his slow-working brain could marvel at the endurance of his own wings.

The Gantlet

THE COMMITTEE ON BIRD PROTECTION *fears that the following species must be placed on the list of those whose survival is doubtful: the California condor (less than fifty birds surviving); ivory-billed woodpecker (less than thirty birds known); Eskimo curlew (population— if any—unknown). . . .*

No additional information on the Eskimo curlew is available. It is of course quite possible that the bird is extinct, but the few reports of its existence during the past decade lead one to hope that it has been merely overlooked by observers. It would seem advisable, however, for the American Ornithologists' Union to attempt to establish relationships with organizations or persons in Argentina and perhaps other South American republics that might be in a position to make investigations. If curlews were found to be wintering in that country it is possible that through the Argentine National Park Directors or other means, some steps might be taken to assure better protection. . . .

Chapter Nine

For nine months of migration each year the curlews were the pawns on a great two-continent chessboard and the players that decided the moves were the cosmic forces of nature and geography—the winds, tides and weather. Winds determined the direction the birds would fly. Tides and rainfall, by controlling the availability of food, determined each flight's goal. Now another player, an ocean current, entered the game.

The Pacific's massive Humboldt current, which sweeps northward from the antarctic along South America's western coast, carries chilled water almost to the equator. The onshore breezes that each afternoon blow in to the Peruvian coast are dry winds for there is little evaporation of moisture from the cold Humboldt water. So the narrow coastal strip between the Andes and the sea is a parched region of sandy desert plateaus where rain rarely falls. Few rivers tumble down the Andes' western slopes into the Pacific to create the estuary mudflats on which the tides can scatter the foodstuffs of the sea for the shorebird flocks. So here the shorebirds eat sparsely. They are tired and thin after the high Andes crossing but the coastal deserts conceived in the Humboldt's antarctic water drive them on without rest.

The curlews followed the narrow Peruvian beaches northward, flying hard each night until the dawn, using every hour of daylight in the wearying search for food. They were always tired and never fully fed. There was neither time nor energy now for the courtship displays, little time even for rest.

In less than a week they covered two thousand miles and reached the sandy flatlands of Punta Parinas near the equator, where the South American coast turns back northeast-ward toward its juncture a thousand miles away with the Isthmus of Panama.

March was almost here. Far to the north, spring would be moving up the Mississippi Valley, greening the cottonwoods and prairie grasses. The curlews were still south of the equator, the tundra was still six thousand miles away. Now the arctic beckoned with a fever and fierceness that their aching and wasted breast muscles couldn't still.

Here the coast swung in a great twenty-five-hundred-mile crescent east, north and west to the rich highlands of Guatemala, but straight north, across the bight of the Pacific enclosed by this crescent, Guatemala was only twelve hundred miles away. The male curlew was still hungry, his crop half-filled, when night began cooling the hot sands of the Parinas desert. He climbed into the tropic twilight and the female followed close behind. And he turned north, away from the low coastland, out into the Pacific where the landfall of Central America lay twenty-four hours of flying away.

They flew silently, wasting neither breath nor energy with calling to each other. It would be an ocean crossing only half as long as the exhausting autumn flight down the Atlantic from Labrador to South America, but the crowberries of

Labrador always assured that the autumn flight could begin with bodies fat and fully nourished. Now they were wasted and thin. In two hours their stomachs were gaunt and empty again.

The southeast trades were left behind after four hours of flying and they entered the calm, windless area of the doldrums at the equator. But the sea below was far from calm. It danced wildly in small, steep waves with foamy, hissing crests—a battleground of waters where the cold Humboldt current met the warm flow of the equatorial current and battled for possession of the sea. Then, even by moonlight, they could see the ocean's color suddenly change as they left the cool green Humboldt waters behind and flew on over the deep blue of the equatorial sea. The air became warmer abruptly.

The moon set and the dawn came. Shortly after dawn they reached the region of northeast trades, a crosswind that made flight easier. But the day rapidly turned hot and the stifling, humid air soon canceled the benefit of the wind.

They flew hour upon hour, the speed of their wingbeats never varying for a moment from the monotonous, grueling three or four beats a second. The glaring sparkle of the sun on the water diminished and finally disappeared as the sun approached its zenith. The sea turned a deeper blue. Then the sun dropped toward the west, the sparkle returned to the wave crests half a mile below and the air grew hotter still. Since the South American coast had disappeared in the darkness of the night before, there had been nothing to break the flat emptiness of sea except an occasional albatross gliding on gigantic, unmoving wings. But the curlews flew northward unerringly, never deviating, their brains tuned more keenly to the earth's direction-giving forces than any compass could be.

The male, partially breaking air for the female, was suffering greater fatigue. The sharp, periodic pains of his breast muscles had changed to a dull, pressing, unabating ache in which he could feel his heart thumping strenuously. He could have obtained some rest by moving back and letting the female lead, but the realization that she was close behind, drawing on the energy of the air that his strength produced, her flight a dependent part of his own, was a warm and exhilarating thrill that made him cling staunchly to the lead position.

The sun went low in the west and his strength dropped to the point where no amount of stubborn mental drive could keep his wings working at the old harrowing pace. But still he clung to the lead. His wingbeat slackened and the flight speed dropped. As soon as she noticed it, the female, who had been silent for almost twenty-four hours, began a low, throaty, courtship *quirking,* and it gave him strength as no food or rest could do. She repeated it at frequent intervals, and the sun dropped close to the horizon sparkling the sea with a million golden jewels of light, and their wings drove them endlessly on.

The sun was setting when the hard blue of the sea at the horizon ahead of them became edged with a narrow, hazy strip of grey-blue. For several minutes it looked like a cloud, then its texture hardened, and behind it higher in the sky emerged the serrated line of the Guatemalan and Honduran mountain ranges. The outline of the distant volcanic peaks sharpened. The lowland close to the sea changed from blue to green, and a white strip of foaming surf took form at its lower edge. There was still a half hour of daylight when the curlews reached the palm-fringed beach. They commenced eating immediately. When darkness came the pain of hunger and fatigue was already diminishing.

They fed busily all next morning, but the feeding was not good for the beaches were scattered and narrow, and swept clean by the Pacific's surf. By noon the day was very hot, but the curlews flew again. They flew inland now, for this was the Central American summer and the grassy highlands of the interior would have a rich crop of grasshoppers. They flew across the coastal plain, which rose gently into the mountains behind. The black fertile volcanic soil was thickly covered with breadfruit trees, coconut palm and plantations of banana trees and sugar cane. In an hour they had climbed a mile above sea level, moving suddenly from tropics to a temperate zone where the air was dry and cool. They climbed higher into mountainous country, then entered a narrow valley that led them through to the rolling tablelands beyond.

They flew four hours and finally landed on a hilly plateau two hundred miles inland from the Pacific. Here, for the first time, the curlews joined the hosts of migrants that were flowing northward to overtake the North American spring. In the forested valleys were swarms of tanagers, thrushes and warblers, all feeding busily to store energy for the long night flights. On the grassy uplands were flocks of other shorebirds and bobolinks. But there was no bird song, for song was the proclamation of the breeding territory, and the breeding territory for most was still two thousand miles away.

On the sloping hills grasshoppers swarmed everywhere. The grass was trampled and cropped close by great herds of sheep, so the insects were easy to find. The curlews fed until their crops and stomachs were gorged. With nightfall thousands of other migrants began passing overhead. Except for the occasional one that passed in silhouette across the face of the moon, they were hidden in the dark, but their lisping chorus of flight notes was an uninterrupted signal of their passage. But the curlews waited, for winter still gripped their arctic nesting grounds and here they could fatten for the final dash north.

They waited a week, feeding well, straggling slowly northward each day. Their bodies grew firm and plump again and with the return of strength the mating urge burned like a fever within them. By the end of the week they had moved out on the Yucatan peninsula to its tip. Five hundred miles northward across the Gulf of Mexico were the swampy shores of Louisiana and Texas, with nothing beyond but the flat unobstructed prairies reaching almost to the arctic.

The Gantlet

. . . But the greatest killings occurred after the birds had crossed the Gulf of Mexico in spring and the great flocks moved northward up the North American plains.

These flocks reminded prairie settlers of the flights of passenger pigeons and the curlews were given the name of prairie pigeons. They contained thousands of individuals and would often form dense masses of birds extending half a mile in length and a hundred yards or more in width. When the flock would alight the birds would cover 40 or 50 acres of ground. During such flights the slaughter was almost unbelievable. Hunters would drive out from Omaha and shoot the birds without mercy until they had literally slaughtered a wagon load of them, the wagons being actually filled, and often with the sideboards on at that. Sometimes when the flight was unusually heavy and the hunters were well supplied with ammunition their wagons were too quickly and easily filled, so whole loads of the birds

*would be dumped on the prairie, their bodies forming
piles as large as a couple of tons of coal, where they
would be allowed to rot while the hunters proceeded to
refill their wagons with fresh victims.*

*The compact flocks and tameness of the birds made
this slaughter possible, and at each shot usually dozens
of the birds would fall. In one specific instance a single
shot from an old muzzle-loading shotgun into a flock of
these curlews, as they veered by the hunter, brought
down 28 birds at once, while for the next half mile every
now and then a fatally wounded bird would drop to the
ground dead. So dense were the flocks when the birds
were turning in their flight that one could scarcely throw
a missile into it without striking a bird. . . .*

*In addition to the numerous gunners who shot these
birds for local consumption or simply for the love of killing, there developed a class of professional market hunters, who made it a business to follow the flights. . . .*

*The field glass was used by the hunters to follow
their flights. . . . There was no difficulty in getting quite
close to the sitting birds, perhaps within 25 or 35 yards,
and when at about this distance the hunters would wait
for them to arise on their feet, which was the signal for*

the first volley of shots. The startled birds would rise and circle about the field a few times, affording ample opportunity for further discharge of the guns, and sometimes would re-alight on the same field, when the attack would be repeated. Mr. Wheeler has killed as many as 37 birds with a pump gun at one rise. Sometimes the bunch would be seen with the glass alighting in a field two or three miles away, when the hunters would at once drive to that field with horse and buggy as rapidly as they could, relocate the birds, get out, and resume the fusillade and slaughter. . . .

In the eighties the Eskimo curlew began decreasing rapidly. . . .

Chapter Ten

March had come. In Canada far to the north, robins, bluebirds and kildeers were already nest building. Here on the Yucatan coast the later migrants seethed with the excitement of the migration time. In winter the over-crowding of the tropics went unnoticed, but when the physiological stirrings of the breeding cycle started, it drove them outward in a frantic search for space where each pair could be alone. Some birds like the swifts and swallows, which could feed on the wing as they flew, migrated by day, leisurely following the Mexican coastline north. But most of the birds dammed up on the Yucatan tip, like a rushing river suddenly blocked, and there they waited, gathering strength, until an evening with favorable wind would send them by thousands into the falling night, out into the wide sweep of the Gulf of Mexico.

One afternoon after two or three days of calm the wind freshened and a restlessness seized the small bird flocks. Bobolinks and thrushes were rising into the air, making short flights out over the water and returning, testing wind and wings. For the curlews, the five-hundred-mile migration across the Gulf would be no more than an average night's flight. But for the smaller songsters, with half the flight speed of the curlews, it was the migration's most rigorous ordeal and the time of starting had to be carefully appraised. By midafternoon many of them were not returning from the test flights. They were climbing high above the surf and in twos and threes were continuing seaward until the black specks of their bodies dissolved into the blue of the sky. By sunset the Yucatan shore was strangely quiet. Only the curlews and a few other shorebirds remained.

It was dark and a full moon was rising when the curlews flew. On the other ocean flights the curlews and golden plovers had been alone, but now they flew an oversea flyway that was dotted with other birds. In two hours the curlews began overtaking the smaller birds that had started earlier. The air was filled with call notes, and wings glistened silver in the moonlight all around. There were easier island routes across the Caribbean and Gulf of Mexico along which the birds could have hopped island to island and rarely been out of sight of land. But the land area of the islands was too small to provide food for the migrant hordes, so most of them followed the main land masses to Yucatan, then crossed the Gulf in a single, nonstop flight. The curlews passed cuckoos that would nest in New England, thrushes and bay-breasted warblers that would mate in the dark spruce forests of the far north, blackpoll warblers that would continue on to Alaska, bobolinks and dickcissels that would fan out across the midcontinental prairies, and brilliant little vermilion flycatchers that would stop and nest as soon as they reached the Louisiana coast. But among the birds that had left Yucatan that afternoon was one species that the curlews never overtook. Many hummingbirds, hardy midgets weighing no more than a tenth of an ounce, had started with the others. But now they were far ahead, outdistancing them all, their tiny wings churning the air with seventy beats a second. Most of the birds would fly twenty hours before they reached the American mainland. The curlews would take ten hours. The hummingbirds would do it in eight.

After four or five hours, the curlews had passed through the flight of smaller birds and were alone again. Suddenly the air grew cool and heavier, giving more lift to their wings, and the easterly trade wind shifted almost to south. Feathery scuds of cloud dimmed the moon at intervals, then the clouds massed into a thick, black, lumpy ceiling and the night was very dark and the gulf waters below were lost in the night's blackness. The wind shifted easterly again, then within fifteen minutes it reversed itself entirely and was blowing from the north. It was a gusty erratic wind. The curlews turned westward to keep it abeam. And then the rain came; it was almost a solid wall of driving water.

After the first explosive outburst, the wind and rain moderated into a steady, lingering storm. It lasted about five hours and the curlews came through the rear of the storm into a clear sky just as the sun was rising. Normally they would have flown on, high over the lagoons and salt marshes of the Texas coast, to land on the prairies of the alluvial plains far inland, but the storm had tired them, their wet wing feathers clung together clammily and responded awkwardly to their pulsing muscles, and the curlews glided low over the beach as soon as they reached the coast.

It was a long, narrow island of sand dunes and grassland that stretched for miles paralleling the coastline. They skimmed across it for a minute or two, seeking a promising spot for feeding before alighting. In the hollows of the sand flats there were numerous ponds with water replenished by the rain of the night before. Hosts of other shorebirds that had left the Yucatan coast ahead of the curlews had also been forced down by the storm and they fed and preened intermittently at the pond edges. The curlews passed over several flocks of plovers and willets, then they breasted a dune and came out suddenly over a broad patchwork of marshy ponds that was dotted with hundreds of Hudsonian curlews. The Hudsonians called to them noisily and the two Eskimo curlews set their wings and pitched down among them.

But they remained with the Hudsonians only that day. At nightfall the two Eskimo curlews flew on alone and in brilliant moonlight two hours later they landed on prairie a hundred miles inland.

Now the migratory restlessness eased again and the curlews were content to wait while the spring moved on ahead of them. Instinct, not reasoning, told them that all the obstacles of the migration were behind and now for three thousand miles to the arctic there were only the great flatlands of the American and Canadian plains, teeming with food, lacking mountains, lacking even a range of hills large enough to interfere with the homecoming flight. It was the home stretch and they could span it in a week if need be. But the migratory urge was temporarily dead. The curlews didn't know that the tundra would not be ready for the nesting for more than two months yet. They only knew that the Texas prairies were rich with the insect life of awakening spring. And they felt an urge to stay.

They waited three weeks, moving barely another hundred miles inland in the whole period. Flocks of the smaller migrants streamed past them overhead almost every night, for they would nest far below the arctic in breeding grounds that were already greening with spring. When the time came, the curlews would span in one night what the small birds covered in three.

It was early April when the restlessness seized them and they began making brief night flights again. They flew easily, stopping always many hours before dawn, sometimes not moving at all for several days at a time. They would wait as the spring moved northward far ahead of them, then in a couple of rapid night flights they would overtake and pass the spring again and wait for it to catch up. The signal to move was the blooming of the willows on the river bottomlands. When the fluffy catkins opened, dusting the evening breezes with the yellow pollen, they would take to the air and fly until they reached a point farther north where the willow buds were unopened and the prairie grass still brown. Then they would wait, feeding luxuriantly on the capsules of grasshopper eggs, which their sensitive bills could feel in the damp soil, and when the willow catkins pierced their buds the curlews would fly northward again.

Each week they moved faster, for the advancing spring picked up speed as it reached more northern latitudes.

The Gantlet

THE AUK

A Quarterly Journal of Ornithology

Published by

THE AMERICAN ORNITHOLOGISTS' UNION

Lancaster, Pa.

General notes. Natural hybrids between Dendroica coronata *and* D. Auduboni. . . . *Rivoli's hummingbird* (Eugenes fulgens) *in Colorado. . . . Eskimo curlew in Texas. Two Eskimo curlews which appeared to be a mated pair were seen in March at Galveston, Texas, by the writer and a number of Houston observers. The birds were amongst a huge assemblage of marsh and shorebirds, including buff-breasted and other sandpipers, blackbellied plovers, eastern and western willets, various herons, and hundreds of Hudsonian curlews. All were feeding over a wide area of sand flats, shallow ponds and grassy patches on Galveston Island, which parallels the coast. Nearness of the Eskimo curlews to*

*Hudsonians gave fine opportunity for comparison.
Smaller size of the Eskimo and shorter length of bill
were obvious, and movements of the birds, in brilliant
midafternoon sunlight, clearly showed the large black
wing area and lack of median head stripe. Fully an
hour was spent checking every identification mark
through eight-power glasses at a range of less than one
hundred yards from our parked car. . . . As is often the
case along the Texas Gulf Coast during spring migration, a heavy rainstorm and change of wind from south
to north during the previous night brought down a
swarming visitation of migrants.—(Sgt.) Joseph M.
Heiser, Jr. . . .*

*A Summary of the Spring Migration.
Undoubtedly the most noteworthy record was the obser-
vation of a pair of Eskimo curlews on Galveston Island,
Texas, the first acceptable record of this species in several years. For twenty years only an occasional lone
Eskimo curlew has been seen and the fact that these
were probably a mated pair makes it a record of great
significance. As long as one pair remains there is hope
that the species may yet escape extinction. . . .*

Chapter Eleven

Now it was corn-planting time on the Nebraska and Dakota prairies and great steel monsters that
roared like the ocean surf were crossing and recrossing the
stubble fields leaving black furrows of fresh-turned soil in or-
derly ranks behind them. Most of the shorebirds shunned the
growling machines and the men who were always riding
them. Yellowlegs and sandpipers would stop their feeding
and watch warily when the plowman was still hundreds of
yards off, then if the great machine came closer they would

take wing, whistling shrilly, and not alight again until they were a mile away. But the Eskimo curlews had little fear. Far back in the species' evolutionary history they had learned that, for them, a highly developed fear was unnecessary. Their wings were strong and their flight so rapid that they could ignore danger until the last moment, escaping fox or hawk easily in a last-second flight. So their fear sense had disappeared, as all unused faculties must, and while other shorebirds relied on wariness and timidity for survival, the Eskimo curlew relied entirely on its strength of wing.

The curlews followed the roaring machines closely, for the white grubs and cutworms that the plows turned up were a rich and abundant food.

All the time their reproductive glands had been swelling in the annual springtime rhythm of development, the development keeping pace with the northward march of spring, so that their bodies and the tundra would become ready simultaneously for the nesting and egg laying. As the physical development came close to the zenith of its cycle, there was an intensification of emotional development too. With high body temperatures and rapid metabolism, every process of living is faster and more intense in birds than any other creature. When the breeding time approaches they court and love with a fervor and passion that matches the intensity of all their other life processes.

Now many times a day the male curlew's mounting emotion boiled over into a frantic display of love. It had become a much more violent display than the earlier acts of courtship. First the male would spring suddenly into the air and hover on quivering wings while he sang the clear, rolling, mating song—a song much more liquid and mellow now than at any other time of year. After a few seconds his wings would beat violently and he would rise almost straight upward, his long legs trailing behind, until he was a couple of hundred feet above the prairie. There he would hover again, singing louder so that bursts of the song would reach the female, bobbing and whistling excitedly far below. Then he would close his wings and dive straight toward her, swerving upward again in the last few feet above her head and landing several yards away.

Panting with emotion, singing in loud bursts, his throat and breast inflated with air and the feathers thrust outward, he would hold his wings extended gracefully over his back until the female invited the climactic approach. She would bob quickly with quivering wings and call with the harsh, food-begging notes of a fledgling bird. Then he would dash

toward her, his wings beating vigorously again so that he was almost walking on air. Their swollen breasts would touch. The male's neck would reach past her own and he would tenderly preen her brown wing feathers with his long bill.

It would last only a few seconds, and the male would dash away again. He would pick up the largest grub he could find and return quickly to the female. Then he would place it gently into her bill. She would swallow it, her throat feathers would suddenly flatten, her wings stop quivering, and the lovemaking abruptly end. For as yet the courtship feeding was the love climax; their bodies were not yet ready for the final act of the mating.

For a couple of hours after each courtship demonstration the passion and tenseness of the approaching mating time would relax, for the love display was a stopgap that satisfied them emotionally while they awaited the time for the physical consummation.

They moved north steadily, a couple of hundred miles each night. The male's sexual development matured first and he was ready for the finalizing of the mating. His passion became a fierce, unconstrainable frenzy and he spent most of each day in violent display before the female. But with each courtship feeding her tenseness suddenly relaxed and the display would end.

It was mid-May and the newly plowed sections of rolling, Canadian prairie steamed in the warming sun. They followed closely behind the big machine with the roar like an ocean surf. The grubs were fat and they twisted convulsively in the few seconds that the sun hit them before the curlews snapped them up. Now the snows of the tundra would be melting. In the ovaries of the female the first of her four developing eggs was ready for the life-giving fertilization.

The male flung himself into the air, his love song wild and vibrant. He hovered high above the black soil of the prairie with its fresh striated pattern of furrows. The roar of the big machine stopped and the curlew hardly noted the change, for his senses were focused on the female quivering excitedly against the dark earth far below. The man on the tractor sat stiffly, his head thrown back, staring upward, his eyes shaded against the sun with one hand. The curlew dove earthward and the female called him stridently. He plucked a grub from the ground and dashed at her, his neck outstretched, wings fluttering vigorously. He saw the man leap down from the tractor seat and run toward a fence where his jacket hung. Normally, at this, even the curlews would have taken wing in alarm, but now the female accepted the courtship feeding and her wings still quivered in a paroxysm of mating passion. She crouched submissively for the copulation and in the ecstasy of the mating they were blind to everything around them.

The thunder burst upon them out of a clear and vivid sky. The roar of it seemed to come from all directions at once. The soil around them was tossed upward in a score of tiny black splashes like water being pelted with hail.

The male flung himself into the air. He flew swiftly, cling-ing close to the ground so that no speed was lost in climbing for height. Then he saw the female wasn't with him. He circled back, *keering* out to her in alarm. Her brown body still crouched on the field where they had been. The male flew down and hovered a few feet above her, calling wildly.

Then the thunder burst a second time and a violent but invisible blow blasted two of the biggest feathers from one of his extended wings. The impact twisted him completely over in midair and he thudded into the earth at the female's side. Terrified and bewildered at a foe that could strike without visible form, he took wing again. Then the bewilderment overcame his terror and he circled back to his mate a second time. Now she was standing, *keering* also in wild panic. Her wings beat futilely several times before she could raise herself slowly into the air. She gained height and flight speed laboriously and the male moved in until he was close beside her.

He continued to call clamorously as he flew, but the female became silent. They flew several minutes and the field with the terrifying sunlight thunder was left far behind. But the female flew slowly. She kept dropping behind and the male would circle back and urge her on with frantic pleas, then he would outdistance her again.

Her flight became slower and clumsy. One wing was beating awkwardly and it kept throwing her off balance. The soft buffy feathers of the breast under the wing were turning

black and wet. She started calling to him again, not the loud calls of alarm but the soft, throaty *quirking* of the love display.

Then she dropped suddenly. Her wings kept fluttering weakly, it was similar to the excited quivering of the mating moment, and her body twisted over and over until it embedded itself in the damp earth below.

The male called wildly for her to follow. The terror of the ground had not yet left him. But the female didn't move. He circled and recircled above and his plaintive cries must have reached her, but she didn't call back.

A long time later he overcame the fear and landed on the ground close to her. He preened her wing feathers softly with his bill. When the night came the lure of the tundra became a stubborn, compelling call, for the time of the nesting was almost upon them. He flew repeatedly, whistling back to her, then returning, but the female wouldn't fly with him. Finally he slept close beside her.

At dawn he hovered high in the grey sky, his lungs swelling with the cadence of his mating song. Now she didn't respond to the offer of courtship feeding. The tundra call was irresistible. He flew again and called once more. Then he leveled off, the rising sun glinted pinkly on his feathers, and he headed north in silence, alone.

The snow-water ponds and the cobblestone bar and the dwarfed willows that stood beside the S twist of the tundra river were unchanged. The curlew was tired from the long flight. But when a golden plover flew close to the territory's boundary he darted madly to the attack. The arctic summer would be short. The territory must be held in readiness for the female his instinct told him soon would come.

Epilogue
Fred Bodsworth

When *Last of the Curlews* was written and first published in 1955, there had not been a report of an Eskimo curlew since the pair on Galveston Island, Texas, referred to in the story, which had been ten years before. It was assumed that the bird had become extinct sometime after 1945.

However, commencing in 1959, and for the following five years, one or two migrating Eskimo curlews were seen again each spring at Galveston. Then in September, 1963, a curlew flying at the head of a flock of shorebirds in Barbados was shot by a hunter. When the hunter saw that his bird was not the familiar whimbrel (the modern name for the Hudsonian curlew), he turned it over to a local rare bird collector, who recognized it as something different and put it in his deep-freeze. It remained there more than a year, and then a Christmas card from the collector to James Bond of the Academy of Natural Sciences in Philadelphia alerted Bond to the possibility it might be an Eskimo curlew. (Yes, Ian Fleming's 007 *was* named after the ornithologist James Bond.) A couple of months later, eighteen months after the bird was shot, Bond visited Barbados, examined the contents of the deepfreeze, and determined that it was an Eskimo curlew. The skin is now in the Academy's collection and it is the last known specimen.

For a short time ornithologists wondered if the Barbados hunting victim was the same curlew that had been turning up on Galveston Island in recent springs, and the last individual of its species. But apparently it was not, for another was seen in Texas in the spring of 1964.

Since then other Eskimo curlews, usually only one or two at a time, have been reported somewhere between the Argentine wintering grounds and arctic Canada every couple of years. *Eskimo Curlew: A Vanishing Species?* by Canadian biologists Bernie Gollop and Tom Barry and Californian curlew historian Eve Iversen, published in 1986 (by the Saskatchewan Natural History Society, Box 1121, Regina, Saskatchewan S4P 3B4), lists twenty-five reports for the forty-one years between 1945 and 1986. And there have been a few more reports continuing into the nineties.

No nest has been found since the 1860s, and no birds showing any breeding behavior have been discovered during nesting season in arctic Canada, despite diligent searching by Canadian biologists.

Undoubtedly some of the Eskimo curlew reports are iden-tification errors. Not only does the whimbrel resemble the Eskimo curlew, but there is also an Asian bird called the little curlew, an Old World version of the Eskimo curlew, that is an even more look-alike species. The little curlew has been identified in California and can be encountered along the North American Pacific Coast as a very rare straggler from Siberia, where it nests. But some of the reports are no doubt valid records of Eskimo curlews. In addition to the Barbados specimen, there are good photos of the Galveston birds of the early 1960s.

So a small population of Eskimo curlews, some guess as few as ten or twenty birds, still lives on, making their long and perilous migration each year. It is conceivable that one Atlantic storm, or one arctic mining development, could destroy them all. But up to now, apparently for about a century, this small dogged remnant has struggled on, still managing to find one another with at least enough regularity each arctic spring to keep their species' gene pool flowing. May those last of the curlews prevail.

March 1995

Afterword

Murray Gell-Mann

Can the human race learn, while there is still time, how to coexist with a great diversity of bird life on this planet? Fred Bodsworth's moving account of the Eskimo Curlew's slide toward extinction raises the issue of how humankind is affecting bird species in general, and, more broadly, how we are decreasing the diversity of plants and animals and whole natural communities in many parts of the world. Above all, it raises the question of how we can acquire the wisdom to become less destructive while supporting our huge—and still growing—human population at a reasonable standard of living, a standard that billions have not even achieved so far.

Seeking greater wisdom includes learning how to be a good ancestor as well as a caring relation—not only of other human beings but also of the other organisms with which we share the biosphere of this planet. After all, we are related to them too, albeit more distantly, as science has revealed since the discovery of biological evolution.

The story of the evolution of all life on Earth involves adjustment to continually changing conditions. Sometimes changes are small enough that individual organisms can accommodate them physiologically or by using inherited instincts or intelligence. Changes that are not so small may be gradual enough that particular species can respond by means of evolution. That process takes place, of course, over many generations, through the natural selection of accidental genetic alterations that tend to make individuals fit better with the new conditions. Even when changes are too severe or too rapid for individuals to adapt or for each species to undergo successful alterations, life can still respond through evolution at the community level, with the extinction of appreciable numbers of species and the eventual appearance of new ones.

In most such cases, any net reduction in species diversity is moderate.

These developments have been going on since life began on Earth some four billion years ago. But every once in a while, at rare intervals, sometimes of tens or even hundreds of millions of years, changes occur that are so great and so quick that none of these levels of adaptation or evolution can prevent a huge wave of extinctions, an upheaval in which biological diversity is greatly reduced over a large part of the biosphere.

For example, it is now believed that the impact on our planet of a large meteorite or small asteroid, which formed the crater of Chicxulub on the shore of the Yucatan peninsula about sixty-five million years ago, contributed significantly to the famous Cretaceous extinction, in which a great many species of animals and plants were destroyed, including most of the dinosaurs. (It appears that at least one group of dinosaurs survived and that birds are their descendants. In that sense, birds are the only remaining dinosaurs.)

Since modern human beings appeared on the scene more than a hundred thousand years ago, we have brought about, at an accelerating pace, the extinction or near-extinction of quite a number of species and sometimes of entire genera or families. In some cases we have virtually wiped out whole ecological systems. Still, these losses, while very considerable, are small compared to those we may induce in the near future if we do not adopt the right policies. By producing rapid and ever more extensive changes in the biosphere—especially in the various natural communities that inhabit it—we are at risk of causing a huge wave of extinctions comparable to the Cretaceous catastrophe, greatly reducing biological diversity on our planet as if we had hit it a gigantic blow. Do we really want to do that?

To some of us it seems obvious that we should try to avoid such a result, that it would be foolish to destroy in a few decades a large part of what biological evolution has achieved over tens of millions of years. A few of us go so far as to say that one especially clever kind of primate has no business wrecking the world in which it and all its relatives have to live.

Others require specific arguments showing how a great loss in biological diversity would cause trouble for us clever primates. There is no lack of such arguments. We all know how interconnected the lives of different organisms are in a natural community. Many of us have heard how the deliberate killing of many of the common Eurasian Tree Sparrows in China led to plagues of troublesome insects. We have heard of the powerful medicines that are discovered as natural substances in our forests, either by exploration or by consulting shamans who have preserved the results of centuries of trial and error by traditional societies. We know that our food plants can often be improved and made more robust by crossing them with related species that grow in the wild. All over the world, people are aware that forests protect us from soil erosion and the deterioration of watersheds. Nowadays it is common knowledge that forests also counteract the greenhouse effect and air pollution; they are sometimes called the lungs of the planet.

Still, the most striking feature of our threat to biological diversity is how little we yet know about what we are destroying. The various natural communities across the world, each with its own large set of organisms interacting in complex and often subtle ways, are still in great part unexplored. Do we really want to degrade them without ever having understood how they are constituted, how they function, and what benefits they may confer on us?

Humans are part of nature too. In most places on Earth people have been present, interacting with the other local organisms, for a very long time, sometimes in rough equilibrium with their natural communities and sometimes causing significant damage, whether in the course of trying to make a living or in more frivolous ways. The present age is special, however, because human population has increased so enormously and because advances in technology have greatly enhanced our capacity to degrade the environment. We are now able to alter rapidly and completely the character of the landscape nearly anywhere on Earth.

We are also producing worldwide effects on air, water, and soil. For example, we are causing substantial changes in the composition of the atmosphere, giving rise to such phenomena as the enhancement of the greenhouse effect and the depletion of the protective ozone layer. Some people demand that scientists furnish absolute proof of the deleterious effects on human life of such phenomena before they will pay for mitigating them, even when the measures contemplated are ones that are desirable for other reasons, such as saving energy. Unfortunately, action must be taken soon if the dangers are really to be reduced, while the proofs can be improved only slowly and can never reach perfection unless and until the predictions are fulfilled.

Do we really want to keep monkeying carelessly with the composition of the atmosphere on the only planet we have, rather than take precautions against the severe consequences that most scientists consider likely? After all, the Western nations during the Cold War took many trillions of dollars worth of precautions against events such as a Soviet attack on Western Europe, not nearly so likely as global warming, but understood to be highly undesirable.

In considering strategies for avoiding a huge wave of extinctions, one may very well ask whether the human race is preparing to do anything about the risk of another asteroid collision. Such events, although rare, are so disastrous that the average rate of damage from that cause is calculated to be comparable with the average rate of damage from earthquakes. Right now intensive discussion is in fact under way about how to employ, for the defense of the planet on behalf of all life, some of the technologies developed over the last few decades for fighting a thermonuclear war.

Of course global effects are not the only ways in which we humans are threatening biological diversity. The human race is applying many other pressures that lead to extinctions and near-extinctions. One obvious example, so well described in this book, involves excessive killing for food and for sport. Significant in the case of the Eskimo Curlew, this has played a role in the destruction of many other kinds of animals around the world, including a number of bird species. But when flying birds are made to "run the gantlet" of hunters, there may be other motivations involved. Sometimes birds are killed to avoid real or imagined destruction of crops or domestic animals.

This was a factor in the disappearance of the Carolina Parakeet, which was, like the Eskimo Curlew, very numerous in the United States in the early nineteenth century. It takes considerable imagination today to picture the landscape of many of our eastern states sprinkled with groups of colorful native parrots, even in the depths of a snowy winter. Those flocks would sometimes descend on orchards and fields and consume considerable quantities of fruit and grain. Angry farmers reacted in a predictable way. Moreover, if one bird was wounded or killed, the rest of the flock would swoop noisily over the victim—not an effective response if the predator is a man with a gun.

The Carolina Parakeet was also hunted for sport, food, and feathers. But it was not only an excess of hunting that exerted severe pressure on the species. Another factor in its disappearance was habitat destruction on a huge scale. Extensive areas of the United States east of the Mississippi that had once been covered with forest were cleared for the planting of crops. Today we might have the wisdom to leave some large areas of forest undisturbed, although unfortunately they might still be cut up into parcels that are too small. We shall return later to the subject of habitat destruction, which is probably the most important threat to bird diversity world-wide.

Huge initial populations offered no guarantee against the extinction of the Carolina Parakeet, any more than for the Eskimo Curlew. Probably both species needed sizable num-bers, gathered in flocks, for survival. So, apparently, did an-other North American species, the Passenger Pigeon, which was even more plentiful. Until 1870 or so, accounts described flocks a mile wide that took hours to pass over, darkening the sky. Only about thirty years later, the species was extinct in the wild, and the last specimen died in captivity in 1914. The destruction of the Passenger Pigeon is attributed partly to the clearing of forests and partly to hunting, sometimes for sport and for the prevention of crop destruction, but mainly for food. As in the case of the Eskimo Curlew on the Great Plains, the harvesting of the Passenger Pigeon in the states east of the Mississippi was a large enterprise. Professional hunters were employed in both cases. John James Audubon described in 1831 what it was like to be in the company of a group of such hunters when the pigeons arrived.

> *Everything was ready, and all eyes were gazing on the clear sky, which appeared in glimpses amidst the tall trees. Suddenly there burst forth a general cry of "Here they come!" The noise which they made, though distant, reminded me of a hard gale at sea passing through the rigging of a close-reefed vessel. As the birds arrived, and passed over me, I felt a current of air that surprised me. Thousands were soon knocked down by the pole men. The birds continued to pour in. The fires were lighted, and a magnificent, as well as wonderful and almost terrifying sight presented itself. The pigeons, arriving by thousands, alighted everywhere, one above another, until solid masses as large as hogsheads, were formed on the branches all round. Here and there the perches gave way under the weight with a crash, and falling to the ground, destroyed hundreds of the birds beneath, forcing down the dense groups with which every stick was loaded. It was a scene of uproar and confusion. I found it quite useless to speak, or even to*

shout to those persons who were nearest to me. Even the reports of the guns were seldom heard, and I was made aware of the firing only by seeing the shooters reloading.

In *Vanished Species,* David Day continues the story of the killings:

Local part-time hunters generally used guns, clubs, poles, smudge pots, and even fire to kill the birds at nesting sites. The best professional hunters used huge, specially designed traps and nets. Some of the large nets were baited with decoy birds. These birds were called "stool pigeons." They were captured birds with their eyes sewn up and their legs pinned to a post or "stool." The fluttering wings of these blind birds attracted other pigeons which were then caught up in the huge nets and slaughtered. Some of the net traps were capable of capturing 2,000 birds at once.

The pressure of the hunters was relentless. As late as 1878, near Petoskey in Michigan, hunters descended on a nesting site [forty miles long and three to ten miles wide]. Their efficiency was astounding . . . .

> *By 1896 there were only 250,000 Passenger Pigeons left. They came together in one last great nesting flock in April of that year outside Bowling Green, Ohio, in the forest on Green River near Mammoth Cave. The telegraph lines notified the hunters and the railways brought them in from all parts. The result was devastating—200,000 carcasses were taken, another 40,000 were mutilated and wasted. 100,000 newborn chicks not yet at the squab state and thus not worth taking were destroyed or abandoned to predators in their nests. Perhaps 5,000 birds escaped.*
>
> *The entire kill of this hunt was to be shipped in boxcars to markets in the east, but there was a derailment on the line on the day of the shipping. The dead birds packed in the boxcars soon began to putrefy under a hot sun. The diligent hunters' efforts were wasted: the rotting carcasses of all 200,000 birds were dumped into a deep ravine a few miles from the railway loading depot.*

Another famous extinction, that of the Great Auk, took place under quite different conditions from those described by Audubon and by David Day, but still resulted, at least in part, from excessive harvesting of an easily available resource. Particular islands and bird rocks scattered through the North Atlantic once harbored large colonies of those flightless birds, skillful swimmers but awkward on land. They were easy prey for sailors, who clubbed them to death in great numbers. For a long time their feathers were in demand, and the bodies could always be used for food and even for fuel, since they were full of oil. During the seventeenth and eighteenth centuries, the population of Great Auks was severely depleted and by the nineteenth century comparatively few of them were left. The last record of a live bird dates from 1844. Toward the end, skin and egg collectors vied with one another to acquire the increasingly scarce specimens. Even scientists participated in this competition. Stevens's Auction Rooms in London, where many disappearing kinds of birds and mammals were featured, had the telegraphic address "Auks, London."

No one made a serious effort to save the species. It was typical of nineteenth-century attitudes toward the reduction of biological diversity—and of human cultural diversity as well—to argue that of course many animals and plants and many traditional human societies would have to disappear to make way for the inevitable march of progress. The losses might be regrettable but those organisms, those tribes were doomed; there was really nothing to be done to prevent their extinction. As Arthur Cleveland Bent (cited by Fred Bodsworth) remarks about the confiding Eskimo Curlews, "no one lifted a finger to protect them until it was too late," although many noticed their decline. Only toward the end of the century did concerned people form conservation organizations, such as the Audubon Society in the United States.

The Audubon Society was founded to fight against such threats to native bird life as the collecting of egret plumes and the like for ladies' hats. (In the late nineteenth century it was even fashionable to wear hats made of a variety of whole, stuffed songbirds, say several kinds of warblers.) The Everglades National Park in Florida was created, in great part, to protect heron species hunted for their plumes. The Audubon Society provided wardens to fend off the plume hunters and a kind of war ensued. Stories abound of hunters tying wardens to Manchineel trees, which secrete a poison that causes powerful skin irritation. Later, the International Council for Bird Preservation was organized and it carried to a successful conclusion the worldwide campaign against commercial plume hunting for the millinery trade. Today, of course, a number of other conservation organizations are flourishing and are engaged in the struggle for the preservation of nature. Those in the United States that have wide-ranging international programs of conservation and associated sustainable development include the World Wildlife Fund, the Nature Conservancy, the Wildlife Conservation Society, and Conservation International.

We have mentioned that in the middle of the nineteenth century the reaction of collectors, even distinguished scientists, to an impending extinction was often to race to obtain a specimen or an egg before the extinction was complete, perhaps disposing of the last living individual in the process. Zoos, too, were sometimes guilty of such practices. Nowadays, however, many zoos have turned into conservation organizations and are engaged in captive breeding of threatened animal species, including birds, and returning members of those species to the wild to strengthen or reestablish populations. Thus the famous New York Zoological Society has become the Wildlife Conservation Society, which sponsors conservation activities in many parts of the world in addition to running the zoos of New York City.

The helplessness of the Great Auk when attacked on land was not an isolated case. Birds that have lived for a long time on an island or an island group uninhabited by human beings or other predatory mammals have in many cases, under conditions of low threat, evolved characteristics that make them exceedingly vulnerable when our species arrives on the scene. They may be flightless and rather unwary, like the Dodo on Mauritius, a tempting target for sailors and colonists after the island was discovered by Europeans at the end of the sixteenth century. Even though its flesh was said to be somewhat disagreeable to eat, the Dodo was pursued with vigor. Within a century, it was extinct.

Some large flightless birds of the southern continents and islands are grouped in the order Ratites (or Struthioniformes). It is believed that these birds have not evolved flightlessness recently but are descended from flightless ancestors that inhabited the ancient continent of Gondwanaland, from which Africa, South America, Australia, Antarctica, and innumerable islands split off many tens of millions of years ago. Among the ratites are the Ostrich of Africa, the Emu of Australia, the cassowaries of Australia and Indonesia, the rheas of South America, and the kiwis of New Zealand, which, although small, have huge eggs.

When the first Polynesian settlers came to New Zealand, they found—in addition to the kiwis—many species of moas, very large ratites, which they hunted for food. The tallest kind reached a height of twelve feet or so. It appears that all the moas have died out. Although climate change may have had something to do with their extinction, human predation surely played a very important role. A few individuals belonging to large species seem to have survived until after the arrival of European colonists. A sighting of one of the smaller species was even claimed in the twentieth century.

The arrival of humans on an uninhabited island can have a devastating effect even when the island is very large, like Madagascar, the "continent island." There, the most massive of the known ratites, the Great Elephantbird, *Aepyornis maximus,* flourished until a few centuries ago. As far as we know, it was the heaviest bird that ever lived. It weighed as much as a thousand pounds and laid eggs about a foot long, fragments of which can still be discovered in certain places on Madagascar. Presumably human activity, after the arrival of the Malagasy peoples, played a role in its demise. Most likely a few elephantbirds were still alive when the earliest French colonists showed up in the seventeenth century, since the first French governor, Etienne de Flacourt, heard of such a bird and gave a brief account of it, as cited in *Vanished Species.*

> *The Vouron Patra is a giant bird that lives in the*
> *country of the Amphatres people and lays eggs like*
> *the Ostrich; so that people of these places may not*
> *catch it, it seeks the loneliest places.*

Flacourt was killed by pirates on the way back to France and never had a chance to elaborate on his description.

On an island previously uninhabited by predatory mammals, vulnerable bird species are threatened not only by newly arrived humans but also by the cats, dogs, rats, and other creatures they bring with them. Take the Stephens Island Wren, which lived on the small island of that name in the Cook Strait between the North and South islands of New Zealand. It was not a true wren but a member of a special New Zealand family. The Stephens Island Wren may have been the only passerine bird entirely incapable of flight. In any case, its flightlessness evolved during its long residence on the island, where there were not enough threats to make the ability to fly worth retaining. Once people settled the island to the extent of building a lighthouse there, the species was both discovered and exterminated by the lighthouse keeper's cat, Tibbles, whose kills supplied the specimens that are found today in several collections around the world.

On some islands, rats accompanying human immigrants have been responsible for exterminating particular kinds of birds. On Guam, it is the accidental importation of the infamous brown tree snake that has wrought havoc with native bird populations. Often a plague of snakes or rats annoys human beings, and mongooses are then imported to reduce their numbers. Unfortunately, the mongooses then go after the native birds as well.

On a great many islands throughout the world, populations of feral pigs, goats, sheep, cats, dogs, rats, and other mammals are causing extensive habitat destruction. Conservation organizations are generally in favor of exterminating these animals or drastically reducing their numbers, and in many places they are taking steps to eradicate them. In some areas, however, the feral mammals are hunted for food or for sport and the hunters are eager to preserve them. An interesting reversal of roles occurs, with conservationists trying to wipe out animals that hunters want to protect.

That has happened in the Hawai'ian Islands, for example. The interior of each of the larger islands contains a great many endemic species of plants and animals—that is, ones found nowhere else in the world. Endemic birds are now concentrated mostly in mountain rainforests, where nearly all the native plants are also endemic. Rooting, browsing, and trampling of vegetation by feral pigs and goats open up paths through those forests that permit introduced plants to spread. The native birds, many of them quite remarkable, are adversely affected by the resulting habitat destruction. Even worse for the birds are the diseases spread by introduced insects, especially the Night Mosquito, which arrived in 1826 and carries avian malaria. That mosquito stays below two thousand feet or so of altitude, and its absence in the highlands is a principal reason some native perching birds survive there.

Plant and animal introductions have seriously affected, in a matter of centuries, the impressive biological diversity of the Hawai'ian Islands. Of course, all the native Hawai'ian organisms are descended from immigrant species that arrived after the islands rose from the sea millions of years ago. But during all that time a rich variety of endemic plants and animals arose there, particularly through the process known as adaptive radiation, in which a single species arriving by accident in an archipelago evolves into a great many daughter species, filling a wide variety of hitherto empty ecological niches. Colonizing a number of islands, where they have evolved in relative isolation from one another, the related forms have often differentiated into separate species, even when they occupy similar niches on different islands. A famous example of Hawai'ian adaptive radiation is that of Drosophilid fruit flies: most of the world's species occur only on the Hawai'ian Islands.

Some of the endemic birds, including a number of kinds of flightless geese, seem to have succumbed after the arrival of the Polynesians. Quite a few of the remaining endemic bird species belong to the Hawai'ian honeycreepers, which form a subfamily of the finch family. In the course of their adaptive radiation they have developed widely different beaks, suitable for feeding on the nectar of variously shaped flowers, on insects, and on a number of kinds of seeds. They also vary a great deal in plumage, and some of the honeycreepers are quite striking in appearance. The 'I'iwi, for example, living in the high forests of the larger islands, has orange-red and black and white feathers and a long, very curved orange-red beak. It is still found in reasonable numbers, but many other Hawai'ian honeycreepers have not been so fortunate. Most species are quite rare and at least seven of them have disappeared altogether since they were first known to science. Introduced diseases of birds are thought to have contributed to these extinctions, along with destruction of habitat and predation by introduced mammals. In 1973, a new species of honeycreeper was identified in a particular patch of forest on Maui and given the freshly coined Hawai'ian name Po'o-uli. It would be a pleasure to report that it is doing well, but sadly that does not seem to be the case. (For one thing, an introduced snail eats native snails on which the Po'o-uli feeds.)

While the honeycreepers are apparently descendants of accidental immigrants from the American side, another group of Hawai'ian land birds belongs to the honeyeater family, which is widespread in Australia and the Southwest Pacific. Of the honeyeaters that used to live in the Hawai'ian Islands, only the Kaua'i 'O'o and the Bishop's 'O'o on Maui seem to have survived into the second half of the twentieth century. Along with the Bishop's 'O'o on Moloka'i and the 'o'o species of Hawai'i and of O'ahu, they were long hunted for their few brightly colored feathers. The O'ahu kind was a royal bird, supplying plumes for the feather cloaks of kings. In principle each captured 'o'o was released after the yellow tuft beneath its wings was plucked, but in practice the birds must often have been eaten instead. It is not considered likely, though, that the use of the honeyeaters by the native Hawai'-ians was enough to cause their undoing. More likely other pressures, including those from introduced mammals and from avian diseases, gave the additional push that drove them all to the brink of extinction.

At one point it was thought probable that all the Hawai'-ian honeyeaters had completely disappeared. Then in 1960 the smallest one, the Kaua'i 'O'o, was rediscovered in the Alaka'i swamp. The number of reported individuals kept dwindling, however, until only a single pair was known. One of that pair then disappeared, and for years the remaining bird kept calling for a mate that never turned up. Finally, the last report of that lonely individual came in the mid-1980s. Thus the Kaua'i species may be extinct. Meanwhile, on Maui, the Bishop's 'O'o was rediscovered in 1981 on the slopes of Haleakala. Perhaps it will survive, although the experience of its relatives is not in the least encouraging.

One might think that bird lovers would always find them-selves on the side of preservation, but sometimes they pre-sent a serious danger instead. Certain species have been driven to extinction or the verge of extinction by collection as cage birds. Many rare parrots, especially macaws, have been victims of this practice. CITES (The Convention on Inter-national Trade in Endangered Species) outlaws international traffic in such birds, but enforcement is difficult and spotty. Spix's Macaw, a beautiful Brazilian species, was reduced, as far as we know, to one individual in the wild and a few dozen in captivity. An attempt is now being made to breed the wild male by releasing in his vicinity a captive female, one that may formerly have been his mate. (At the time of the release, he had been associating with a female of another species, but the female Spix's Macaw seems to have taken her place.) There is, of course, only a small chance that a wild population can be restored by such means, and even if it can, the genetic diversity within that population will presumably be low. Still, the experiment is worth trying.

In other cases, foreign birds have been deliberately introduced into the wild, sometimes for sentimental reasons. In one remarkable instance, a group of eccentric bird-lovers formed the "American Acclimatization Society," based on the idea that the English-speaking countries of North America should have all the different kinds of wild birds mentioned in the works of Shakespeare. Of the resulting importations from Britain, that of the European Starling has produced a permanent effect. Since its successful introduction to Central Park, New York, in 1890 it has become fantastically numerous on the mainland of the United States and in Canada, displacing native birds on a significant scale. The House Sparrow had already been introduced in several places, starting in the 1850s; one reason for bringing it over was to counteract plagues of dropworms that disturbed such mid-nineteenth century events as the Easter Parade on Twenty-third Street in New York. Both birds are now real nuisances, just like introduced weeds such as kudzu. (Think, too, of the Walking Catfish from Southeast Asia, which scrambles from one body of water to another in Florida, greatly reducing the numbers of native fish wherever it goes.)

Nostalgia allied to the Shakespeare phenomenon led to the importation to New Zealand of numerous British birds, which are now more conspicuous in many places than the native ones. In the Hawai'ian Islands, the land birds of lower elevations now belong mostly to introduced species such as the Common Myna. That situation, as already mentioned, is at least partly attributable to avian malaria, but the disease probably arrived with some of the foreign birds, which are much less sensitive to it than are the native species.

For all the problems caused by unrestrained hunting, intentional and unintentional introductions of foreign organisms, and trapping of wild birds to be caged, the greatest threat to bird diversity is surely the destruction of habitat. Even the Eskimo Curlew may have been a victim of habitat destruction as well as deliberate killing. This book describes the arrival of the lone male curlew and its Golden Plover companions on the Argentine pampas after their long and difficult migration from the arctic. They gorge on the insect life of the plains and marshes while they molt and recover their strength. But in 1891 the Argentine supply of grain from Russia was drastically reduced by a crop failure there and the pampas were quickly brought under cultivation without setting aside a large undisturbed portion. That change in the land where the Eskimo Curlew spent the southern spring probably made a great difference to its prospects for survival. Near the other end of the migration route, the plowing of the North American prairies may have had a significant effect as well.

All over the world, ecological systems have been and are still being degraded or fragmented or destroyed. The variety exhibited by those systems is extraordinary. We can easily think of a great many different categories, such as desert, grassland, scrubland, temperate rainforest, temperate montane coniferous forest, temperate deciduous or mixed forest, tropical rainforest, tropical dry forest, tropical montane forest at various altitudes, and tundra. But we should remember that within every category the geographical variation is impressive—so many different kinds of desert, of grassland, of tropical rainforest, and so forth. Each has its characteristic set of organisms, from trees and mammals to microscopic life. In every case the fate of the local birds is interwoven with that of the rest of the system. That is true of the resident bird species and also of those that merely summer or winter or pass through on migration.

We are putting enormous pressure on a large fraction of these ecological systems. Our rapidly growing human population includes many who are very poor and have to scrabble for a living in any way they can, unable to pay much attention to the future. Some others are comparatively well off and consuming considerable resources per person. Both groups, along with those in between, are contributing to ecological stress. But this stress could be reduced in most cases if actions were more carefully thought out by leaders and decision makers and if the issues were better understood by the public.

In many places the gross destruction of tropical forests rich in biological diversity has actually been encouraged by international lenders such as the World Bank. Native forests full of species found only in the local region have been deliberately clear-cut to give way to monocultures of Monterey

Pine or of a particular kind of eucalyptus. Many bird species, such as the Sri Lanka Woodpigeon, are threatened by this kind of substitution. The monotonous plantation does not support them the way the forest did.

If you hire a car at the international airport at Nadi on the Fijian island of Viti Levu, you can drive up into what was once a very rich Fijian tropical forest. Now you will find instead stretches of naked ground alternating with pure pine plantations nearly barren of other organisms. Near the capital of Suva on the same island, you can see by comparison some surviving patches of the original forest, the home of endemic birds such as the Golden Dove and Blue-crested Flycatcher. If it is necessary to exploit the forest economically, wouldn't it be worthwhile to try harder to find markets for highly useful select native woods and for nontimber forest products? The lazy way is to destroy a complex natural heritage, evolved over many millions of years, simply because the native trees can be sold off, say for disposable chopsticks, and a timber market is already known to exist for a simple substitute introduced from abroad. In some places, it is profitable to leave the most of the forest undisturbed and earn money through "ecotourism," which allows visitors from afar to appreciate the wonders of the local ecological system.

Some bird species depend, for critical contributions to their food supply, on the flowering or fruiting of particular plants. Such a plant may have a patchy distribution. Moreover, it may flower only infrequently, perhaps even at irregular intervals depending on the location. Suitable food may thus be abundant at some times and places and very scarce at others. Birds dependent on such a resource are particularly vulnerable to habitat destruction. If they are forest dwellers, incomplete but widespread clearing of their native woods can expose them to the danger of extinction. Their vulnerability should be a signal to us that we are interfering with a very delicate system.

Consider the Purple-winged Ground-dove, a bird of the disappearing Atlantic Forest that used to cover southeastern Brazil, eastern Paraguay, and northeast Argentina. This dove is, at least in part, a bamboo-flower follower. It feasts on the seeds of certain species of bamboo, traveling around to the areas where they have recently flowered. It is intolerant of deforestation and now very rare over most of its former range. A number of bird species in various parts of the world are true bamboo specialists, utterly dependent on the patchy distribution in space and time of particular kinds of flowering bamboo. Encroachment on their habitat increases the distance and the time interval between one episode of gorging and the next, making it difficult for the birds to survive.

It seems that acorns and other mast formed an important part of the Passenger Pigeon's diet. The progressive clearing of forests made it more and more difficult for the pigeon flocks to find large patches of mast, and some ornithologists now assign to this phenomenon a large share of the blame for the extinction of the species.

The threat posed by partial deforestation is not restricted to specialist birds. The breaking up of forests into small parcels, even when the total area remains substantial, causes very serious problems for many forest organisms. One trouble is that most of the forest is then near an edge, and the interaction with the world outside is much more intense. For example, predators can gain access to most places instead of only a few. So can parasites, not just internal parasites like the ones causing avian malaria in Hawai'i, but external ones such as the Brown-headed Cowbird in North America. That species, like the famous Eurasian Cuckoo, lays its eggs in the nests of songbirds, which brood them, along with their own eggs, until they hatch. The cowbird chicks are bigger and more aggressive than the ones that belong in the nest. The young cowbirds take a huge share of the available food and may expel from the nest eggs or young of the host species. An important factor in the decline of the songbird population in North America over the last few decades is increased cowbird parasitism in fragmented woodlands.

The intense selection pressures that human activity imposes on other organisms favor a few kinds, ones that are especially resourceful or adaptable (omnivorous, for example) and ones that tolerate human presence or actually gain benefit from what people do. Take large-scale commercial fishing, for instance, with its enormous discharge into the ocean of tissues from fish and other marine animals. Certain ocean birds follow the fishing vessels and gain a bonanza from what is thrown away. Thus the Northern Fulmar has experienced a significant increase in numbers in recent years.

In the United States, Northern Cardinals and Northern Mockingbirds have extended their ranges considerably since I was a child. A winter cardinal was a great rarity in New York at that time, whereas now it is a common sight far up into New England. Bird feeders are no doubt partially responsible, and perhaps also the rising average temperatures created by urban and suburban development, to say nothing of global warming.

In the Southwest, where I now live, the Common Raven has become very abundant. Pairs of ravens, mated for life, are to be seen in all directions. Is the availability of road kill responsible? While it is encouraging that some avian species are on the increase, we must understand that the balance of nature can be disturbed in that way too. The huge population of ravens may threaten desert organisms on which they prey, such as the Desert Tortoise and the Horned Lizard (often called the Horned Toad).

Sometimes rare birds are endangered by the expansion in numbers of predatory birds that thrive on human settlement. On the island of Sakhalin in the Russian Far East, for example, crows are now very much more common than they used to be, posing a threat to the few remaining Spotted Greenshanks in the area.

Conservation practices can change the pressures that human activities exert on bird populations. Instead of continuing to favor just a few species while reducing the numbers of the rest, we can reverse the decline of hitherto unfavored ones. In the United States and Canada, during the twentieth century and especially the last few decades, some encouraging progress has been made in preservation and even restoration of habitat, careful and integrated pest management instead of reckless application of dangerous pesticides, restrictions on the importation of exotic flora and fauna, and protection of the native birds themselves. In the case of game birds, reasonable regulations for hunting have in many cases been created and enforced.

A number of bird species have recovered that were apparently on their way out at the beginning of the century. Wild turkeys were rapidly vanishing a hundred years ago and are now flourishing in many places. Of course they are popular game birds, but some wild species (such as the Osprey) with no powerful constituency to support them have also recovered as a result of conservation efforts. However, a great deal remains to be done. The overall trend is still toward a reduction of biological diversity, largely as a result of unsustainable economic activity. This is especially true in tropical countries, but holds for temperate countries as well.

To appreciate the importance of acting now to prevent the impoverishment of our environment, including our native birds, imagine a future in which most ecological communities have been eliminated or degraded, in which diversity has been drastically reduced. In that kind of world, what would the bird life look like in the United States and Canada? We would still have the introduced House Sparrows, European Starlings, and Rock Doves. In suitable places, there would be Herring Gulls, Turkey Vultures, American Crows, and Common Ravens. Around houses we would see the usual Blue Jays or Scrub Jays, American Robins, Northern Mockingbirds, Northern Cardinals, or House Finches. But there would be very few warblers, orioles, tanagers, or vireos— very few woodland songbirds, in fact.

Peter Matthiessen has described his view of some of the birds that would remain:

> *In cities where all of the forests have been cut down,*
> *what do you find? You have all these requiem birds:*
> *starlings, blackbirds, vultures, ravens. All these birds*
> *that adapt so well to civilization—harsh-voiced,*
> *gloomy birds feeding on carrion and garbage.*

The whole natural system would be greatly reduced in complexity, greatly impoverished. And for what? To make room for still more people? At any given level of conventional material prosperity per person, our environmental impact is roughly proportional to the population. Yet many discussions of population concentrate on how many human beings the Earth can support. Why would we want to squeeze onto the surface of our planet as many people as possible when we see the extent to which numbers bring so many problems? Why should we not be concerned about the swarming of people into congested megacities that barely function, loss of freedom as more and more regulation of human behavior becomes necessary, high rates of unemployment for those lacking advanced education or special skills, and encroachment of agriculture onto previously wild lands that supported high biological diversity? Why squander quality of life for the sake of mere numbers of humans?

Population is not the only important variable, however. It is also important to cultivate practices that reduce our impact on the environment. We need incentives to adjust aspects of our economic activities, especially our technologies, so that we all live to a much greater extent on nature's income rather than nature's capital. That will be facilitated if our society makes further attempts to charge the true costs of goods and services, including environmental costs, and to reward true benefits. Industry and agriculture will then be increasingly motivated to use gentler technologies and our economy will come much closer to being sustainable.

Although incentives are usually preferable to regulation by micromanagement, some regulation is needed. One aspect of regulation that is often challenged is the protection of endangered species and subspecies. In the midst of arguments about real or imagined conflict with property rights, the main point about endangered organisms is often lost. Those disappearing animals and plants tell us when a given natural system has been so abused that we should protect what is left of it.

Endangered or threatened species or subspecies thus act as surrogates for their natural communities and guardians of ecological systems. Their impending extinction indicates that a whole system has been overexploited and needs a rest. The Northern Spotted Owl is thus protecting the old-growth forests of the U.S. Pacific Northwest. Where those forests have been clear-cut, the tree farms that sometimes replace them cannot support all the organisms that thrive in the forest. Even when substantial areas of old growth are saved, those are inadequate if they are divided up into small patches.

The little Snail Darter, the endangered fish that held up the construction of the Telleco Dam, was telling us that nearly all the tributaries of the Tennessee River had been dammed and that it was time to set aside some of these waters and allow them to run free. When Snail Darters were found in another stream, that was an indication that perhaps a few minor undammed tributaries might suffice as a refuge.

The Red-cockaded Woodpecker is a key element in the ecology of the old-growth Longleaf Pine forests of the southeastern United States. It makes its holes in the trunks of living trees where the wood has been softened by the red heart fungus. The holes are then used by other forest creatures. It does best in unmanaged fire-climax forest, where natural fires are allowed to occur, leaving the older trees alive. But most of the habitat is being managed for timber in such a way that fires are controlled, the "over-mature" trees are cut, the fungus is treated as a pest, and the Red-cockaded Woodpecker is disappearing. The bird is telling us to leave a big chunk of its habitat alone.

In the Southwest, streamside vegetation has been destroyed in many places (sometimes by federal programs for "phreatophyte control," which means getting rid of vegetation that might compete with agriculture for water). The Southwestern Willow Flycatcher is sending us the message (as is the Least Bell's Vireo in California) that we should save what is left of our riparian willow and cottonwood groves.

Often, when a bird species (or another kind of organism)

is endangered because a whole natural community is in trouble, the issue of preservation is stated as one of biological diversity versus jobs for humans. But in many of these cases, the economic activity in question involves precisely the destruction of that natural community, and the jobs will disappear anyway when it is exhausted. If the felling of the remaining old-growth forests of the Pacific Northwest is permitted, it may provide jobs for a few more years, but the employment will not be sustainable. The search for more sustainable alternatives might as well start sooner rather than later. In fact, many loggers in that area have already found satisfactory employment in other sectors.

A few endangered birds in the United States are the beneficiaries of heroic programs to keep them from becoming totally extinct. Those include spectacular birds such as the Whooping Crane and the California Condor, which is now entirely dependent on captive breeding and release into the wild. Is this kind of program worth the cost? I believe so, not just because it is an inspiring public enterprise to keep such national (and international) treasures alive, but also because the effort keeps us aware of how much cheaper and better it is to take protective measures early, when there is more to save and the saving is easier.

In the case of migratory birds, even heroic measures in the summer breeding area are not sufficient, since the birds or their habitat may be destroyed in the wintering area or along the migration route, as in the case of the Eskimo Curlew.

Kirtland's Warbler nests in Jack Pines of a particular height located in a tiny area of the Lower Peninsula of Michigan. There a vigorous conservation program has been in effect, including careful measures for cowbird suppression, but for a long time the Kirtland's Warbler population remained stable rather than increasing. A suggested explanation was that the winter habitat in the Turks and Caicos Islands in the Caribbean might be shrinking. This kind of connection is common and illustrates how important it is for conservationists in North America to help the tropical countries of the Western Hemisphere to carry out conservation and sustainable development activities. In the particular case of the Kirtland's Warbler, it seems to have turned out that the winter habitat is still in good shape, and recently the number counted in Michigan has begun to increase.

International agreements and informal cooperation across national boundaries play a most important role in the efforts to prevent the impoverishment of the fauna and flora of the world. This is particularly true in the case of migratory birds. In fact, some migratory bird treaties have been in place for a long time. The work of the International Crane Foundation, especially in Asia, presents a remarkable contemporary example of how even countries that are not on particularly good terms with one another can be induced to cooperate in the preservation of bird species. Of course people have been especially willing to help save cranes because of their grace and beauty, their huge size, and the veneration with which they have been traditionally regarded in many societies. We need to marshal support not only for such impressive species but also for ones that are much less conspicuous.

We conservationists must often remind people that when a species is extinct it is gone forever, an irreparable loss to the planet. That argument is certainly correct today, but it might not always hold. Progress in biotechnology might make it possible someday to reconstruct species that are long gone by collecting the DNA of ancient, even fossilized specimens, as in the novel *Jurassic Park* by Michael Crichton and the film made from it. Scientists could even begin now to collect DNA from species that are currently threatened or endangered, so that they could be reconstructed more readily after extinction if and when the technology becomes available.

Gregory Benford, a physics professor and science fiction writer, has proposed that such a program be started right away.

The speculation that such reconstruction might be possible in the distant future is, however, no argument against conservation today. It would never be easy to re-create a lost species, and it would be even harder to re-create or repair a whole natural community, with all its organisms connected by a complex web of relationships. But that is the real point of conservation: to save as much as possible of the diversity of such communities, including their threatened and endangered species and much else besides. After all, those communities are critical parts of the ecological systems that support the human economy.

Those who are skeptical of the need for protection of the environment frequently cite the comparatively small fraction of plant and animal species that have definitely disappeared during the last few centuries. (For example, the number of bird species known to be completely exterminated since 1700 is about a hundred.) They miss the point that many species die a lingering death, taking many decades to pass from near-extinction to a situation in which we can be certain that no more individuals are alive. (Indeed, in some cases we may never be completely certain.) For some purposes, extinction and the verge of extinction should be lumped together. We should remember that biological diversity includes not only diversity of species but adequate numbers of each species as well. When that is taken into account, the reduction in biological diversity since 1700 is much more striking.

While it cannot be excluded that a few Eskimo Curlews still survive, it is very doubtful that a viable breeding population exists anywhere. Likewise a tiny number of Ivory-billed Woodpeckers might conceivably linger on somewhere in the United States, perhaps in the Big Thicket wilderness or nearby, although the patch of woods where the last confirmed sighting took place has been chopped down. Every once in a while someone discovers a remnant population of a species or subspecies thought to be extinct. But usually the prospects for long-term survival of such a population are not very bright.

Whether the Eskimo Curlew has really vanished completely or not, Fred Bodsworth's book shows us clearly the flaw in how it was perceived. People must have thought (or rather felt) that it represented an infinite resource rather than merely a rich one. The same must have been true for the Passenger Pigeon, which was probably the most successful bird on Earth until a massive attack was mounted against it by human beings.

Today we face the same problem in a much more general context. Even if we treat our fellow organisms and the biosphere merely as resources, we are mistaken if we regard such planetary resources as inexhaustible rather than finite. Moreover, a dogma taught by many economists only reinforces that misconception—the dogma that by an infinite sequence of substitutions of resources, an infinite chain of technological fixes, each making up for some of the damage caused by the previous one, we can somehow go on forever abusing the planet we inhabit. A related argument runs that when a resource is nearly exhausted, its price rises, a cheaper substitute is then found, and the last little bit of the original resource is then not used. This hardly applies to biological populations, especially when considerable numbers are required for survival, as appears to have been the case for the Passenger Pigeon and the Eskimo Curlew.

The human race must get used to the simple idea that the Earth is really finite. The sooner that occurs, the happier the outcome will be.

Recommended Reading

Collar, N. J., M. J. Crosby, and A. J. Sattersfield. *Birds to Watch 2: The World List of Threatened Birds.* BirdLife Conservation Series 4. Cambridge, England: BirdLife International, 1994.

Collar, N. J., and S. N. Stuart. *Threatened Birds of Africa and Related Islands: The ICBP/IUCN Red Data Book.* 3d ed., pt. 1. Cambridge, England: International Council for Bird Preservation, and International Union for the Conservation of Nature and Natural Resources, 1985.

Collar, N. J., et al. *Threatened Birds of the Americas: The ICBP/IUCN Red Data Book.* 3d ed., pt. 2. Cambridge, England: International Council for Bird Preservation, 1992.

Day, David. *Vanished Species.* New York: Gallery Books, 1989.

Diamond, Antony W., Rudolf L. Schreiber, Walter Cronkite, and Roger Tory Peterson. *Save the Birds.* Boston: Houghton Mifflin, 1989.

Ehrlich, Paul R., David S. Dobkin, and Darrel Wheye. *Birds in Jeopardy.* Stanford: Stanford University Press, 1992.

Fuller, Errol. *Extinct Birds*. New York and Oxford: Facts on File Publications, 1987.

King, W. B. *Endangered Birds of the World: The ICBP Bird Red Data Book*. Washington, D. C.: Smithsonian Institution Press in cooperation with the International Council for Bird Preservation, 1981.